Framing the Farm Bill

Framing the Farm Bill

INTERESTS, IDEOLOGY, AND THE AGRICULTURAL ACT OF 2014

Christopher Bosso

University Press of Kansas

Published by the University Press of Kansas (Lawrence, Kansas 66045), which was organized by the Kansas Board of Regents and is operated and funded by Emporia State University, Fort Hays State University, Kansas State University, Pittsburg State University, the University of Kansas, and Wichita State University

Library of Congress Cataloging-in-Publication Data

Names: Bosso, Christopher J. (Christopher John), 1956– author.
Title: Framing the farm bill : interests, ideology, and the Agricultural Act of 2014 / Christopher Bosso.
Description: Lawrence, Kansas : University Press of Kansas, 2017. | Includes bibliographical references and index.
Identifiers: LCCN 2016047592
ISBN 9780700624195 (cloth : alk. paper)
ISBN 9780700624201 (pbk. : alk. paper)
ISBN 9780700624218 (ebook)
Subjects: LCSH: United States. Agricultural Act of 2014. | Agricultural laws and legislation—United States. | Farm law—United States.
Classification: LCC KF1681.A3282014 B67 2017 | DDC 343.7307/6—dc23
LC record available at https://lccn.loc.gov/2016047592.

British Library Cataloguing-in-Publication Data is available.

Printed in the United States of America
10 9 8 7 6 5 4 3 2 1

The paper used in this publication is recycled and contains 30 percent postconsumer waste. It is acid free and meets the minimum requirements of the American National Standard for Permanence of Paper for Printed Library Materials Z39.48-1992.

CONTENTS

ACKNOWLEDGMENTS

First, thanks to Will Masters of the Tufts University Friedman School of Nutrition Science and Policy, where I spent a spring 2014 sabbatical offering a graduate seminar on the politics of food and starting work on a book about food policy. Will, my host and, at the time, chair of the Department of Food and Nutrition Policy, asked me to give a talk to the larger Friedman community about the recently enacted Agricultural Act of 2014. "Sure," I replied, and then spent a few weeks poking around the legislative history and votes. Regardless of how the talk turned out (it's on iTunes), I got hooked and threw out the other project I had been working on. So if you don't like this book, blame Will Masters. Also, thanks to my Friedman School colleagues Tim Griffin, Nichole Tichenor, and, especially, Parke Wilde, whose *Food Policy in the United States* (Routledge, 2013) was an essential source.

This book also was inspired in part by the memory of William P. Browne (1945–2005) of Central Michigan University, the rare political scientist who studied agricultural policy and politics. The University Press of Kansas published two of Bill's most influential works—*Private Interests, Public Policy, and American Agriculture* (1988) and *Cultivating Congress: Constituents, Issues, and Interests in Agricultural Policymaking* (1995)—so having Kansas publish this book seems fitting. Bill and I crossed paths a few times in my early years as a scholar, and I remember his graciousness and advice. I will never match his depth of understanding about farm and rural policy (the two are not necessarily the same), but I hope this book speaks well of his legacy.

At Northeastern University, my longtime home, thanks to my many colleagues and students who suffered through my incessant yammering about the Farm Bill, which otherwise might still be a mystery to most folks at my East Coast urban school. Northeastern also has given me the space to experiment with new courses and try new research directions, a flexibility we scholars must never take for granted. Thanks also to Northeastern graduate students Armin

Akhavan, Skye Moret-Ferguson, and, especially, Claudia Larson for their research and editorial support.

Unlike other "how a bill becomes a law" tales, this book is based entirely on publicly available materials, not insider knowledge, so I owe a deep debt to the small cadre of journalists who cover agricultural policy and politics. In this regard, a special thanks to Jerry Hagstrom, the most knowledgeable of them all, both for his exhaustive coverage over many years and for taking the time to talk with me once I had read everything he wrote. I would also be remiss if I didn't mention the work of David Rogers (*Politico*), Ron Nixon (*New York Times*), and other contributors to the few nonspecialist national publications that still pay attention to the Farm Bill. If, as the saying goes, journalism is history's first draft, I was well served.

Thanks, of course, to Chuck Myers and the other good folks at the University Press of Kansas and to the anonymous reviewers, whose input I tried to incorporate. Any errors, factual or interpretive, are mine alone, which always irks me. But there you go.

Finally—as always—thanks to Marcia. So much of our time together seems to be marked by book acknowledgments, so you must be my muse.

1

What's Going on in Kansas?

The First Kansas is one of the nation's most rural House districts. In its current form, the "Big First" takes up more than half the state, stretching from the outer suburbs of Topeka in the northeast over 300 miles westward along the Nebraska border to Colorado and then curling south another 200 miles around the Fourth District to the Oklahoma state line. At 57,000 square miles—equal to the entire state of Illinois—it is also one of the nation's largest House districts. To put it in perspective, the Thirteenth New York, the nation's smallest House district, starts at 100th Street in Manhattan and heads north all of eight miles through Harlem along the Hudson River to Van Cortland Park. Both districts have around 720,000 residents, as per the House apportionment rules based on the 2010 census. There the similarities end. Where the Thirteenth New York is urban (70,000 people per square mile), ethnically and racially diverse, younger, poorer, and Democratic, the First Kansas is rural (*11* people per square mile), 90 percent white, older, more affluent, and Republican. The Thirteenth New York is Harlem, Columbia University, and the Bronx. The First Kansas is Fort Hays, Kansas State University (in Manhattan, Kansas), and Dodge City.

More to the point of our story, the First Kansas is agriculture: vast stretches of wheat and grain sorghum (used as cattle feed)—an average farm size of 1,000 acres (or 1.5 square miles)—interspersed with "concentrated feeding operations" where thousands of head of beef cattle are fattened for slaughterhouses operated by Cargill and National Beef using largely Hispanic immigrant labor.[1] While manufacturing and the service sectors are now greater overall drivers of the state's economy, agriculture remains important, and the Big First is Kansas agriculture's epicenter.

Not surprisingly, those representing the First Kansas in Congress, including Representative (later Senator) Robert Dole and Representative (later Senator) Patrick Roberts, were devoted to and left indelible fingerprints on agricultural policy, and they made sure that Kansas got its fair share of federal funds to support farming. Dole, as we will later see, also played a key role in expanding federal nutrition programs serving low-income Americans, while Roberts made his mark by promoting reforms in federal farm programs. In historical terms, then, representatives from the Big First have played pivotal roles in developing the nation's farm and food programs.

In January 2011 the First got a new representative, Tim Huelskamp, who filled a seat left empty when incumbent and fellow Republican Jerry Moran ran successfully for the Senate seat vacated when another Republican, Sam Brownback, became governor. (Note the pattern: whoever represents the Big First in the US House, a safe Republican seat, invariably becomes a top contender when one of the state's US Senate seats becomes available.) Huelskamp, a state senator, won the seat after beating five other Republicans in a hotly contested primary, after which he achieved an easy victory in the general election.

Huelskamp grew up on a family farm in Fowler, Kansas (2010 population, 590), and like many future members of Congress, he developed an early fascination with politics and public policy. In 1995 he returned to his southwestern Kansas hometown after finishing his doctorate in political science—not the normal law school path taken by most members of Congress—and immediately jumped into politics. A year later he was elected to the state senate—one of the youngest state senators in decades. He would be reelected three times with ease, based largely on his core conservative values on issues such as abortion and same-sex marriage, as well as his deeply held view that government had become too big and too expensive. Perhaps reflecting his early training in a Catholic seminary, or maybe just because he is contrarian by nature, Huelskamp also displayed a willingness to criticize his Republican colleagues if he felt they weren't doing enough to pursue conservative goals. In fact, in 2003 he was removed from the state senate's key Ways and Means Committee for clashing once too often with party leaders. His reputation with voters for being uncompromising in defense of his—and their—values, aided by financial support from national conservative advocacy groups, enabled Huelskamp to claim the Big First's open seat in 2010 as part of the Tea Party wave that put Republicans back in control of the House of Representatives.[2]

Befitting his district's agrarian status, Huelskamp promptly claimed the Big First's "traditional" seat on the House Agriculture Committee, maintaining a lineage established by Dole and Roberts in particular. However, anyone who thought that Huelskamp had gone to Washington just to promote Kansas agriculture was soon disabused of that notion. To the surprise of no one who had paid attention to his state senate career, Huelskamp fast became a vocal thorn in the side of House Speaker John Boehner and other Republican leaders who, in his opinion, weren't carrying out voters' wishes. He also refused to compromise on cutting federal spending, even when the all-important Farm Bill was up for reauthorization. In fact, despite pleas from his state's leading agricultural industries and farm groups to support its passage, Huelskamp and a cadre of fellow conservative House Republicans blocked action on the Farm Bill throughout the 112th Congress (2011–2012). In December 2012 Speaker Boehner, furious at Huelskamp's refusal to fall in line with party leaders on key votes, booted him from the important Budget Committee and, to ensure that the lesson hit closer to home, the Agriculture Committee.[3] The Big First now had no seat on House Agriculture for the first time in recorded history. If Huelskamp was shaken by his punishment, he didn't show it. "The Kansans who sent me to Washington did so to change the way things are done," he declared, "not to provide cover for establishment Republicans who only give lip service to conservative principles."[4]

What Is This Farm Bill about Which You Speak?

This book was inspired by a roll-call vote. On January 29, 2014, after three years of sharp ideological and partisan conflict, the US House of Representatives approved the House-Senate conference committee report on the Agricultural Act of 2014 (H.R. 2642), known colloquially as the Farm Bill (as are all versions of this legislation).[5] The final vote (table 1.1) was telling in itself, suggesting significant lingering unhappiness with the final package.

Of interest was the vote's breakdown. The bundle of programs that make

Table 1.1 Final House Vote on H.R. 2642 (January 29, 2014)

	Yea	Nay	Not Voting
Republicans	162	63	6
Democrats	89	103	8
Totals	251	166	14

up any version of the Farm Bill is the foundation of the nation's food production system, and the law typically comes up for renewal every five years. Whatever happens in the course of any single Farm Bill's journey through the legislative process, final passage is practically assured because, as congressional scholar David Mayhew surmises, the compromise package satisfies most legislators' policy, constituency, and, of course, reelection needs.[6] Indeed, the previous edition of the Farm Bill was enacted in 2008 by overwhelming bipartisan majorities—*twice* over vetoes by President George W. Bush.

Not so in 2014: House Democrats split almost evenly on the bill's final passage, forcing Republicans in the majority to overcome defections by one-third of their members. What piqued my curiosity was that two-thirds of the Republicans voting against passage had not been in the House in 2008. Most were so-called Tea Party conservatives hailing from largely suburban and rural districts in the Midwest, South, and Southwest and who, since the 2010 midterm elections, had reshaped the House Republican majority and the party overall.[7] For their part, Democrats who voted against the bill were largely from liberal, urban districts on the East and West Coasts. What was it about the Farm Bill this time around that provoked such clear ideological, partisan, and even intraparty splits?

Another compelling fact was this: while most midwestern farm-state Republicans voted for the final package, which renewed key agricultural support programs through fiscal year 2018 (or September 30, 2018), notable in their adamant opposition were all four House members from *Kansas*—America's breadbasket, a place so synonymous with farming that a sheaf of wheat adorns the official state seal. How is it possible, I wondered, that all four House members from Kansas could be so hostile to the Farm Bill—the *Farm* Bill!—to vote no, even though a yes vote would have been an easy gesture to the folks back home who wanted them to support it? With all due respect to Mayhew, weren't the four of them acting against their own electoral self-interests? Others noticed too: reporters for the *New York Times* observed that it was the first time in recorded history that the entire Kansas House delegation had voted against the final version of a Farm Bill, despite public pleas by the state's agricultural leaders to support passage.[8]

As you may already surmise, one of these four Kansans stood out. Tim Huelskamp's ouster from the Agriculture Committee and his consistent opposition to the Agricultural Act should have hurt his reelection prospects, an outcome that John Boehner likely would have welcomed in the hope that voters in the Big First would elect a more agreeable Republican. But they didn't: Huelskamp would survive a primary challenge and go on to win easy reelection in November 2014. So did his three compatriots—who together made up the nation's most conservative House delegation[9]—no doubt because they adhered to the set of values that got them elected to Congress in the first place.[10]

An intriguing (to me, at least) side note: Huelskamp's story is more interesting, perhaps ironic, to anyone who studies Congress because in 1995 he earned a doctorate in political science and wrote his dissertation on changes in the composition, structure, and institutional roles of the House and Senate Committees on Agriculture from the early 1970s through the mid-1990s.[11] That two decades later he got kicked off the House Agriculture Committee for insufficient loyalty to party leaders is the stuff of (not very good) fiction.

To recap: all four House members from Kansas voted against the Farm Bill, despite pleas from their state's agricultural establishment to support it. All four were reelected anyway. What is going on here?

As is often the case, the central story of the Agricultural Act of 2014 turned out to be about something other than farming. It was about *food stamps* or, to be precise, about the Supplemental Nutrition Assistance Program (SNAP), the nation's primary tool to help low-income Americans buy food. Spending on SNAP had risen sharply since 2008, in large part because of the Great Recession, but also because Congress had expanded program eligibility during the previous decade, so that it now accounted for nearly 80 percent of the Farm Bill's annual costs (figure 1.1).

H.R. 2642 as enacted contained $8 billion in cuts—or savings, depending on one's viewpoint—in SNAP through 2024, most in the form of tighter eligibility requirements. Although the reductions came to less than 1 percent of annual SNAP outlays, they prompted opposition from almost half the House

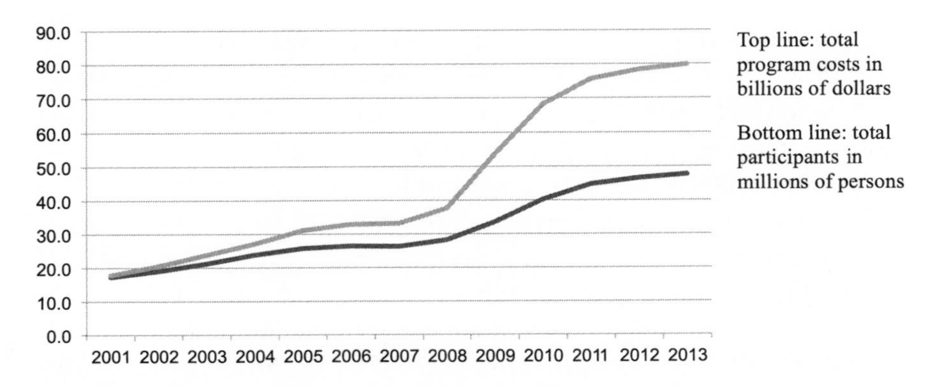

Figure 1.1 SNAP Costs and Enrollment Growth, Fiscal Years 2001–2013
Source: Food and Nutrition Services, US Department of Agriculture, "Supplemental Nutrition Assistance Program Participation and Costs," http://fns.usda.gov/pd/SNAPsummary.htm.

Democrats and a third of House Republicans—for opposite reasons. As Rosa DeLauro, a Democrat from Connecticut, railed that the final version was "nothing more than Reverse Robin Hood legislation that steals food from the poor in favor of crop subsidies for the rich," Marlin Stutzman, a Republican from Indiana, lamented that its passage only perpetuated "the unholy alliance between food stamps and farm programs."[12] In short, liberals like DeLauro voted no to protest cuts in SNAP spending, while conservatives like Stutzman voted no to protest the insufficiency of those same cuts in a program they thought should not be part of the Farm Bill anyway.

Wait, you might be thinking: why are food stamps included in the Farm Bill? That's a story in itself, and it's the central irony of the fight over the Agricultural Act of 2014. Here's the short version: For decades, going back to the early 1970s, food stamps and other nutrition programs (such as providing surplus commodities to food pantries) were included in a series of federal Farm Bills *precisely* to build coalitions of support. Urban liberals who might not care about or might even oppose federal subsidies for a comparatively small group of farmers would support the Farm Bill in return for votes from rural conservatives who might otherwise oppose the politically unpopular food stamp program. In fact, Huelskamp's Big First predecessor, Robert Dole, was instrumental in expanding food stamp coverage while serving in the Senate, in part to cement that urban-rural partnership. However, that cross-ideological, bipartisan coalition, born out of pure political necessity, had frayed by 2011 when Congress took up Farm Bill reauthorization, and by 2014, SNAP was a point of division, not cohesion. How that became the case after decades in which the "farm programs + food stamps" deal had provided mutual benefits, and why House members from Kansas could oppose the Farm Bill and not suffer for their purported sins, is the focus of this book.

Why Follow the Farm Bill?

The path taken by the Agricultural Act of 2014 tells us a lot about the politics of food, and possibly a great deal more about contemporary American politics. Not that we lack for good general-audience books on the politics of food, a topic of interest to many more of us every day. Indeed, any short list includes Eric Schlosser's *Fast Food Nation*, Marion Nestle's *Food Politics*, Michael Pollan's *The Omnivore's Dilemma*, and Daniel Imhoff's *Food Fight*, to name a few.[13] Most of these works are by journalists, nutritionists, and food advocates, and all express strong points of view about changes needed in the food system. There is nothing wrong with that, of course. Some of the most compelling insights into our politics come from advocates seeking to change it. Think about Upton Sinclair's semifictionalized exposé of the mistreatment of workers in Chicago slaughterhouses in *The Jungle* (1906). While Sinclair hoped (in vain) that his book would inspire readers to overthrow capitalism, public uproar about conditions in the slaughterhouses did spur President Theodore Roosevelt and Congress to enact the first federal Meat Inspection Act and the first Pure Food and Drug Act. Or, as Sinclair later famously mused, "I aimed at the public's heart, and by accident I hit it in the stomach."[14]

Curiously absent from any list of recent authors addressing this topic are political scientists, whom one would expect to pay particular attention to the politics of food production and consumption. Indeed, go back a few decades, and you would have no trouble finding political scientists who used agriculture as a lens to examine the broader dynamics of the vote-gathering role of political parties, interest-group organization and political power, bureaucratic culture and routines, and the impacts (good and otherwise) of federal farm policy. Notable among them were Schattschneider, *Politics, Pressures, and the Tariff* (1935); Freeman, *The Political Process* (1955); Bauer,

Poole, and Dexter, *American Business and Public Policy* (1963); McConnell, *Private Power and American Democracy* (1966); Redford, *Democracy in the Administrative State* (1969); and, more recently, Berry, *Feeding Hungry People* (1984); Browne, *Private Interests, Public Policy, and American Agriculture* (1988); and Hansen, *Gaining Access: Congress and the Farm Lobby* (1991).[15]

Today, with a few exceptions, agriculture seems to be overlooked in political science.[16] One reason may be that it is no longer a key economic sector. In the late 1940s agriculture by itself (versus food processing, retail, and restaurants) accounted for more than 20 percent of the nation's economic output and employed more than 20 percent of its workforce. Today, the sector accounts for less than 5 percent of the nation's output and employs less than 10 percent of its workers.[17] So while agriculture remains essential in some areas of the country, its relative centrality to the nation's economy has waned. It might make more sense to focus on health care or online commerce if one wants to study politics and policymaking.

A second reason is that few members of Congress today hail from or depend on the votes of farming areas. This is particularly true for the House of Representatives, whose structure of representation reflects where Americans live. Even in the late 1950s, more than 200 of the 435 House districts were classified as "rural," and many members came from and depended on the votes of farming communities. This created an identifiable cross-party "farm bloc" that had leverage in Congress and in national politics overall.[18] Farmers and the many organized groups representing agriculture mattered, and as such, they were duly studied by scholars. Today, less than 2 percent of Americans are farmers, and only 34 House districts are considered rural.[19] The congressional farm bloc simply shrank as the nation's center of demographic and political gravity shifted to the suburbs and, most recently, to the exurban stretches of home developments, service industries, and chain retail and dining establishments, particularly in the booming Sun Belt.[20] Where people—and political power—go, so goes the attention of scholars.

Finally, agriculture once seemed to represent "normal" American politics. That is, agricultural policymaking typically was organized around specific crops and products, and as such, it reflected the local biases inherent in an institutional structure in which legislators give priority to their respective geographic constituencies over any directives from congressional leaders or the president.[21] If you were a representative from the Big First, you represented wheat. Agriculture also reflected the vote-gathering role of regionally based political parties, which prospered not because of any overarching philosophy about government but because of their ability to pull together winning coalitions in every presidential election. Just as James Madison envisioned when he warned about the "mischiefs of faction" in *Federalist 10*, but political scientist Robert Dahl almost two centuries later depicted as "normal" pluralist politics,[22] agriculture typified the clout of active and intense organized interests—again, think Kansas wheat growers—over an inattentive and unorganized public, whether depicted as voters or consumers. Agriculture was synonymous with mainstream political science conceptions of a client-centered "distributive" politics, with members of Congress ensuring their reelection by being responsive to narrow, geographically specific, organized commodity interests—southern cotton, midwestern corn, western beef—each of which received its respective piece of the Farm Bill pie within the legislative context of formal and informal vote trading. Each Farm Bill was an exercise in distributing just enough benefits to enough distinct constituencies to ensure its passage, including, when their votes were needed, members representing urban food stamp recipients.[23] In the end, almost everyone, save those most concerned about the federal budget or "good" public policy, left happy—or at least not unhappy—with the results.

But on the surface, at least, something had changed. What else could explain how four House members from Kansas could vote against the Farm Bill and survive? Why were there such sharp ideological and partisan divisions over legislation that, on the surface at least, traditionally seemed to sidestep them? Didn't "distributive" politics work anymore? Clearly, our conventional understanding of how the world works needed some updating. In my mind, at least, it was time to revisit the politics of the Agricultural Act of 2014, both because its saga tells us something about current American politics in general and because it is a good story in its own right.

What do such stories tell us? The usual answers come to mind: the "textbook" model of how a bill becomes a law doesn't suffice (if it ever did) to capture the fluidity and complexity of congressional action; following an actual piece of legislation tells us a lot about larger social and political trends and the issues at play; and it is just plain fun, at least for us policy and politics wonks. Equally important, compared with many policy areas, the president is a comparative bystander in Farm Bill politics. That's worth pondering at a time when images of presidential power, or at least centrality, have been (again) overmagnified during post-9/11 and Great Recession–era crisis politics. The Agricultural Act of 2014 wasn't the 2002 Patriot Act. It wasn't even the 2010 Affordable Care Act. But it *was* highly partisan and ideological, reflecting larger political battles that, in the end, made final enactment less certain than at any time in decades.

The Farm Bill is also about more than agriculture. According to Michael Pollan, trenchant critic of the dominant food system, we should call it the "Food Bill," since its many provisions shape food production and, subsequently, how we eat. For example, why have commodity crops such as corn, wheat, and soybeans historically received generous federal government support, while most vegetables and fruits have not? What is and isn't considered "organic"? Why are many agricultural activities exempt from federal environmental laws? Why are food stamps part of the Farm Bill? And who decides all this?[24]

The Farm Bill is also, in the end, about nutrition. As noted, nearly 80 percent of the annual spending under Farm Bills passed in 2002 and 2008 was for the Supplemental Nutrition Assistance Program. How did this come to be? That's a long story examined in more detail in chapter 4, but suffice it to say that decisions made more than forty years ago to put the food stamp program into the Farm Bill, and under the formal jurisdiction of House and Senate Committees on Agriculture, reflected rational calculations by representatives of a fast-shrinking farm bloc. They needed to recruit votes from their urban colleagues for crop-support programs assisting an ever smaller slice of the population. As such, food stamps became the glue that held the Farm Bill together through thick and thin, with rural representatives supporting the program's expansion as their own numbers continued to shrink. They did whatever it took to get enough votes to pass the bill.

Today, most of the Farm Bill's spending goes toward what many Americans see (and not sympathetically) as a handout to people "not like themselves." This makes the battle as much about class, race, immigration, the "deserving poor," individual moral responsibility, and the proper role of government as about growing or eating food. This is why the Farm Bill is about more than agriculture—and why we should pay attention to it.

The Story to Come

This book looks back at the Farm Bill's most recent reauthorization to gain deeper insights into the politics of food and American politics more broadly. Like any narrative in which the legislative process is the central thread, the story follows a generally chronological path. Having said this, it is not a day-by-day, blow-by-blow description of how a bill became a law. Instead, the meandering trail taken by the Agricultural Act of 2014 is used as an organizing device to frame an examination of larger themes in American politics.

Along the way, two stories take center stage. First, we'll follow two Kansans, Senator Patrick Roberts and Representative Tim Huelskamp, previous and current representatives of the Big First, to observe the contrast (and frequent clash) between the former's more traditional approach to balancing competing interests and the latter's arguably more absolutist perspective in defending core ideological values. The focus on Roberts and Huelskamp also highlights the larger changes and struggles within the Republican Party, perhaps *the* defining political story of the current era.[25]

The second story is about food stamps—for decades, the glue that held together that tenuous coalition of rural and urban interests each time the Farm Bill came up for reauthorization. As we will see, the fight over SNAP would be the defining battle of the Agricultural Act of 2014, and it threatened to disrupt long-standing cross-party partnerships and established ways of doing business. The SNAP controversy also shed light on deeper partisan and, especially, ideological disagreements over the scope and direction of federal spending. As such, the debate over the Farm Bill was not simply about farm and food programs but about the role of government itself. That's why the Farm Bill matters to all of us, not just to farmers and food policy advocates.

Chapter 2 reviews the food system and why government policy is so critical to it. The food system did not just create itself; it is shaped by government policy, and for good reason. Chapter 3 looks back as the House and Senate Committees on Agriculture took up the task of reauthorizing the Food, Conservation, and Energy Act of 2008, whose core provisions would lapse unless formally reauthorized by September 30, 2012. This journey back into history helps us understand the origins of US agricultural policy and why, when Congress took up the Farm Bill in early 2011, so many considerations reflected decisions made years, even decades, earlier. We also meet Senator

Patrick Roberts of Kansas, whose career in Congress reflected and contributed to many of the policies being revisited.

Chapter 4 looks at the range of organized interests positioning themselves as Congress prepared to move on reauthorization. Once, these interests would have been limited to farmers' organizations like the American Farm Bureau Federation or commodity groups like the National Corn Growers Association. Today, they include a diverse array of farming, agribusiness, nutrition, environmental, international aid, trade, and consumer interests—each aligned in loose coalitions of support for the issues or programs under consideration, and each brought into the Farm Bill "tent" to guarantee its passage. A major part of that effort was the deliberate decision by rural legislators, typified by Senator Robert Dole of Kansas, to bring food stamps and other nutrition programs under that tent, in no small part to maintain the support of urban America for a shrinking farming population.

In chapter 5 we find the House and Senate Agriculture Committees compelled to adapt to a dramatically changed political dynamic as the Republicans, energized by the conservative Tea Party surge of 2010, gain a majority in the House and set out to slash federal spending. To do so, House leaders shifted power to the chamber's Budget Committee, which ordered major and immediate cuts in agriculture and nutrition programs and imposed limits on the Agriculture Committee's spending. Perhaps more important, broader partisan and ideological conflicts over the budget, including threats to shut down the federal government, caused concern that key farm programs would not be renewed by the deadline. In the middle of this fight, largely among House Republicans, were Tim Huelskamp and other newly elected conservatives determined to rein in the size and power of the federal government, including, if need be, farm programs that were important to the folks back home.

Of particular importance in this regard was the long-standing linkage between commodity and nutrition programs. As detailed in chapter 6, this link endured in the Democratic-controlled Senate, despite efforts by conservative Republicans to narrow program eligibility and rein in costs. Indeed, the Senate acted relatively quickly in mid-2012, precisely because senators from farming states were able to pull together a bipartisan coalition sufficient to overcome opposition by their most conservative colleagues. By contrast, as explored in chapter 7, leaders of the House Committee on Agriculture struggled to get their version of the Farm Bill to the chamber floor, largely because of resistance by Republicans like Huelskamp and because legislators from farming districts had become a distinct minority in the lower chamber. The 2008 Farm Bill formally lapsed on October 1, 2012, only to be extended until September 30, 2013, as part of a late-year deal between Republican leaders and President Obama.

Chapter 8 finds us in mid-2013: Obama won reelection, Republicans maintained control of the House, and Huelskamp, having crossed swords once too often with Speaker Boehner, lost his seat on the Agriculture Committee, prompting talk of a primary challenge in 2014 out of concerns that their feud would hurt Kansas agriculture. The Senate, still controlled by Democrats, again approved a "standard" Farm Bill, including commodity and nutrition programs, but the House remained locked in a fight over SNAP. The House Agriculture Committee succeeded in getting its bill to the floor, only to see it defeated in a shocking final vote as both liberal Democrats and conservative Republicans bailed out, for opposite reasons. How did this happen? House leaders, desperate to show progress, eventually overrode the Agriculture Committee and put the farm and nutrition programs into separate bills, both of which passed in partisan floor votes. Having disconnected the traditional "farm programs + food stamps" linkage, the question was whether either could survive on its own.

Chapter 9 brings us to the House-Senate conference committee in late 2013. The task was now to align the various bills into a single version that both chambers could support. Meanwhile, Congress and President Obama were locked in yet another deadlock over the budget, resulting in a two-week shutdown of the federal government. Here we follow the role of key senators in defending "traditional" agricultural needs, even if doing so required support for nutrition programs that were unpopular back home. We also review the impacts of institutional arrangements such as the conference committee, which allowed defenders of the longtime "farm programs + food stamps" arrangement to regain the advantage and shape the final bill.

We end our story at Michigan State University, where President Obama formally signed the Agricultural Act into law on February 7, 2014.[26] Notable by their absence, despite being invited, were *any* Republican members of Congress. Even at the moment of its signing, passage of the Farm Bill said volumes about the current state of American politics.

2

The Food System

Or, Why Governments Don't Leave Agriculture to the Marketplace

Eating is an agricultural act.
—Wendell Berry

In 1971 President Richard Nixon nominated eminent Purdue University agricultural economist Earl Butz to head the US Department of Agriculture (USDA). Butz, who had served as assistant secretary of agriculture in the Eisenhower administration, faced questions during Senate confirmation hearings about his ties to agribusiness firms such as Ralston Purina, on whose board he sat. Such companies were essential, Butz declared, as "the bushel of wheat in Kansas has no value until it becomes bread for Mrs. Housewife." Even so, Butz promised, he would be a "vigorous spokesman" for all of agriculture, especially farmers.[1]

Butz gained confirmation, and in his five years in office he shaped a US agricultural policy that encouraged, if not outright compelled, the unlimited production of basic commodities such as corn, wheat, soybeans, cattle, and pigs. "Get big or get out," Butz was known to say about (if not directly to) farmers.[2] Under Butz, and to this day, US agricultural policy would be guided by a single goal: plentiful, inexpensive food for all Americans.

That goal, enshrined in successive versions of the Farm Bill, largely achieved its objectives. US agricultural output skyrocketed over the next four decades, even as fewer Americans worked the land. Farms grew larger and became more productive as farmers tapped new seed and animal varieties, more sophisticated farm machinery, chemical inputs ranging from herbicides to animal antibiotics, and technologies such as genetically modified (GM) seeds. Despite mounting criticism about food production's economic, health, and ecological impacts, Butz strongly defended the system he had helped shape, observing in 1998: "We feed ourselves in this country now with about 11 percent of our take-home pay, which means we feed ourselves with about 6 percent of our gross domestic product. That includes all the built in service you get at the store now. Imagine: 11 percent of take-home pay, leaves 89 percent for everything that makes life so wonderful in America."[3] Two decades later, Americans spend around 10 percent of their net income on food, whether eaten at home or elsewhere, the lowest level per capita in the world.[4]

It bears reminding: we Americans are accustomed to plentiful, diverse, convenient, and inexpensive food. To put it plainly, for most Americans, the food system gives us what we want, when we want it, and at a price we're willing to pay—whether at Whole Foods or Walmart. In terms of human history, this is no small achievement. While the United States started out with advantages in arable land, water, and favorable growing conditions, today's abundance is also due to key characteristics of the food system as it evolved from the middle of the twentieth century.

Scale. A sector that was once marked by millions of small farms gave way in the post–World War II period to far fewer yet larger operations (figure 2.1). Farm consolidation was already a fact by the time Butz took over at the USDA: the total number of farms had plummeted from 6.5 million in 1935 to 2.7 million in 1975. At the same time, the average farm size more than doubled, from 155 to 391 acres. Despite a recent modest uptick in the number of farms smaller than 10 acres due to consumer demand for local food, the overall number of farms continues to fall—to about 2.1 million in 2012, with an average farm size of 434 acres.[5] More telling, by 2012, 3.8 percent of farms accounted for 66 percent of the market value of all agricultural products.[6] In this case, scale matters. As one study put it in 2002, "An economically viable crop/livestock operation in the Corn Belt would have between 2,000 and 3,000 acres of row crops and between 500 and 600 sows."[7] In farming, smaller is not necessarily better.

Efficiency. The food system is efficient in classic industrial terms. For example, in 1970 approximately 67 million acres were dedicated to field corn

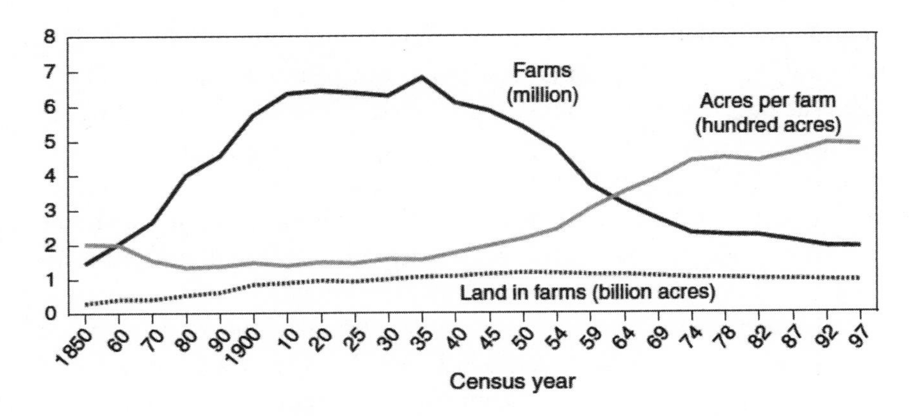

Figure 2.1 Farm Numbers and Size, 1850–1997
Source: David E. Banker and James M. McDonald, eds., *Structural and Financial Characteristics of U.S. Farms: 2004 Family Farm Report,* Agricultural Information Bulletin 797 (Economic Research Service, USDA, March 2005), 5.

(versus sweet corn, which accounts for less than 1 percent of the corn produced), with each acre yielding 72 bushels on average. By 2013, partly due to federal farm policies stressing production and congressional mandates supporting the blending of "renewable" forms of ethanol in gasoline, the amount of land devoted to corn had gone up to 95 million acres. More important, production per acre had *doubled,* to 160 bushels—or about 14 billion bushels overall. Put into perspective, US farmers produce 2 bushels (or 120 pounds) of corn for each of the 7 billion persons in the entire world. About 40 percent of that output goes to corn-based ethanol, and another 40 percent goes into animal feed.[8] What Michael Pollan vividly describes as a "river of corn"[9] enables the United States to produce around 43 billion pounds of poultry, 26 billion pounds of beef, and 23 billion pounds of pork each year—or 290 pounds per American.[10]

Another example is dairy (figure 2.2). From 1970 to 2012, the number of dairy cows in the United States dropped from 12 million to just over 9 million, and the number of dairy operations decreased from 650,000 to about 90,000, due in part to federal milk pricing rules that encouraged producers to be larger and more efficient. The rules worked as planned: average herd sizes grew from 20 to 100 cows, and milk output per cow increased from 9,700 to 21,700 pounds per year. The result was a near *doubling* of produc-

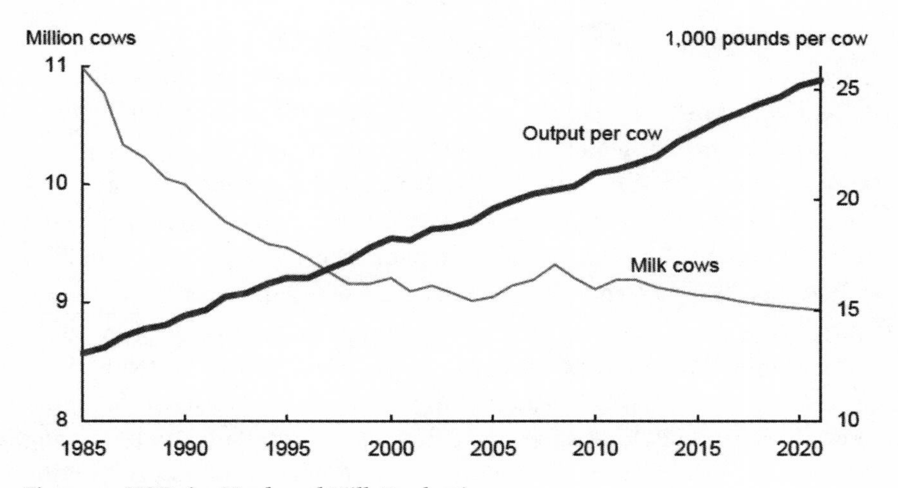

Figure 2.2 US Dairy Herds and Milk Production
Source: James McBride et al., *Profits, Costs, and the Changing Structure of Dairy Farming*, ERS Report 47 (Economic Research Service, USDA, September 2007), 2, http://www.ers.usda.gov/media/188030/err47_1_.pdf.

tion—from 117 billion pounds in 1970 to over 200 billion pounds in 2012.[11] Fewer farms, fewer cows—more milk than ever.

Specialization. Scale and efficiency demand specialization. Whereas farmers once rotated two or three row crops annually, while also raising some cattle and hogs, today's "factory in the field" stresses *monoculture*—devoting all acreage to one crop, such as corn or soybeans. Although farmers may alternate—growing corn one year and soybeans the next, depending on global market demand—they typically grow *all* corn or *all* soy. Specialization also extends to animals, most notably in concentrated animal feeding operations (CAFOs), in which hundreds or even thousands of cattle or hogs and tens of thousands of chickens or turkeys are fattened for market in controlled environments where they consume nutritionally calibrated corn- or soy-based feed.[12] In Iowa, the center of the Midwest's corn and pork production, the percentage of farmers raising hogs fell from 70 percent in 1964 to 11 percent in 2000, yet *total* hog production remained stable.[13] Nationally, the number of farms raising hogs dropped from 700,000 in 1980 to 68,300 in 2012, and facilities with more than 5,000 head now produce 62 percent of all hogs.[14] Similar trends are seen for beef and poultry.

Technology. Agriculture is a temple to technology: large GPS-guided seeders (to optimize planting) and harvesters, computer-calibrated application of fertilizer, genetically modified seed, animal growth hormones and antibiotics, and synthetic chemical herbicides, pesticides, and fungicides. Farmers are even using camera-equipped drones to monitor their far-flung operations. Such capital-intensive technologies have replaced people, enabling the few who farm to obtain higher yields with less labor. In 1940 nearly 31 million Americans, or 25 percent of the population, worked the farm. In 2012 there were 3.1 million—a farming 1 percent feeding everyone else.[15]

Energy dependence. Industrial-scale mechanized, technologized agriculture depends on energy. In addition to the mostly carbon-based fuels that power the nation's electrical grid and move food from farms to processing facilities to consumers, agriculture depends on natural gas to produce the ammonium nitrate that is the building block of synthetic nitrogen fertilizers, petroleum to catalyze a wide range of chemical pesticides, and gasoline and diesel fuel to power tractors, combines, trucks, irrigation pumps, and on-site processing facilities. Total energy use in agriculture comes to about 1.6 trillion British thermal units (Btus) per year, compared with 8 trillion Btus used annually by all US households.[16] Although agriculture has become more efficient in terms of gallons of oil used per bushel of crops produced, the economic health of this sector, like that of the nation, is affected by the availability and cost of oil and natural gas.[17]

Globalization. Ralston Purina (on whose board Butz once sat) is now part of Nestlé, the Swiss-based conglomerate whose operations span the globe, as do those of major US-based food companies Cargill, ConAgra, and Heinz. For their part, US farmers produce far more food than Americans can eat, and agricultural exports amount to more than $140 billion a year, about 10 percent of total US exports by value.[18] Yet Americans also import 40 percent of their fresh fruits and vegetables and 85 percent of all seafood.[19] If you live in Boston and want fresh blueberries in February, you can get them shipped in by air from Chile. What you want, when you want it, at a price you're willing to pay.

Why Governments Manage Agriculture

This food system did not emerge spontaneously or without help. Instead, like most areas of the economy, its composition reflects policies enacted by government, often as a result of the peculiar economics of agriculture itself.

In theory, agriculture behaves like any market. There are sellers and there are buyers. Sellers compete for market share, using product differentiation, price, and service to entice buyers away from other sellers. Buyers, for their part, make choices based on those same factors. Everyone adjusts as new information about product and price becomes available. So, for example, in the smart-phone market, a consumer first decides on the operating system: Apple iOS or Google Android? Choose the former, and an iPhone is the only option. Product differentiation is by core platform. Choose the latter, and the choices expand, with various Android makers competing over product features and price.

Consumer demand for a product in a properly functioning market is *elastic*—that is, affected by price. For example, after Apple unveiled the iPhone in 2007, smart-phone sales boomed as unit prices fell and more (and more user-friendly) options to the iPhone became available. In theory, demand could soften and sales could drop if smart-phone prices spiked, perhaps because of production backlogs or a shortage of an essential material, or if consumer purchasing power dropped due to job losses or other economic factors that made new phones luxuries rather than necessities. The broader point is that product demand is affected by price and consumers' ability to buy.

Most critical, according to the standard view, is that product differentiation plus elasticity of demand should lead to market conditions in which little government intervention is needed, except perhaps to prevent competition-killing monopolies (as in the 1980s, when the federal government broke up the Bell system into competing telephone companies), to allocate "scarce" public resources (such as cell phone frequencies), or to address any environmental, health, and safety side effects of market production, consumption, and waste.[20] However, as the financial crisis of 2008 underscored, markets *do* fail, prompting demands for government intervention to reestablish stable relationships between buyers and sellers.[21]

If markets in general often do not work according to theory, agricultural markets seem to work least well.[22] And by *agriculture*, I don't mean consumer products like the burger you ate at Wendy's or the Old El Paso taco kit you bought at Costco. I mean basic commodities like corn, soybeans, wheat, pigs, beef cattle, poultry, milk, and so on—the raw bulk elements of the food system. It turns out that markets for basic commodities don't work like they do for smart phones. For example, save for organic variants, feed corn is feed corn and soybeans are soybeans. It's not like making a choice between an Apple iPhone and a Samsung Galaxy, or even between a Galaxy and a Motorola Droid. As a result, those producing or buying agricultural commodities typically care most about volume and unit price. It is all about efficiency and marginal cost.

Moreover, in this market, we "eaters" aren't the primary customers. Corn growers sell their entire crops to aggregators like Cargill and Archers Daniel Midland, which convert the raw corn into more refined products (cornstarch, cornmeal, high-fructose corn syrup), which processors like Nabisco and Kraft transform into the more "value-added" products (breakfast cereals, crackers, taco kits, frozen pizza) found on our grocery store shelves. Or, if a farmer raises cattle, the whole herd goes to Tyson or ConAgra for slaughter and processing, where the bulk beef is broken down into more finished products for your local TGI Fridays or supermarket meat case. So Earl Butz had a point about the need for companies to turn raw commodities into finished products. More important, unless you are that nearly mythical creature who cooks everything from scratch using crops you've grown yourself, you raise and slaughter your own meat, and you never eat out, your "choices" are at the tail end of a global food system in which most of the consequential decisions have already been made.

In this regard, the enduring problem for American agriculture is elasticity of demand. We may not *need* an iPhone, but we *do* need to eat. So there is a minimum guaranteed market for the nation's farmers. And, except in rare instances—such as a sharp rise in egg prices in mid-2015 after an avian flu outbreak in the Midwest forced producers to kill millions of hens—higher commodity prices don't necessarily affect consumption.[23] Fortunately, for most of us, a 10 percent increase in bulk dairy prices won't have much of an impact on our milk drinking. But on the flip side, a 10 percent decrease in bulk dairy prices won't induce us to consume more dairy products either. Even we Americans can eat only so much ice cream.

So the trick in any agricultural sector is to produce enough to meet demand while keeping prices high enough to allow farmers to cover their costs and make a profit. This is where the dilemma of imperfect elasticity of demand comes into play. Individual farmers are economically rational insofar that each needs to produce as much corn, soybeans, pork, or chicken as possible to maximize net revenues. Yet, as individual businesses, a single farm is too small to have real leverage in a global marketplace, where competitors are just as likely to be in Brazil as in the next county. Even the largest American or Brazilian soybean grower has nowhere near the global market power that Apple, or even Samsung, enjoys when it comes to smart phones. Moreover, Apple or Samsung can always respond in a timely manner to sagging demand by resetting prices, changing product specifications, introducing a new model, or pausing production. For farmers, however, *not* producing because commodity prices are low is impracticable because of their need to cover "sunk" costs in land, buildings, seed (or animal equivalents), and technology incurred months or years before harvest. A crop planted in April at one global price still has to be harvested in September, whatever the current price. Worse, for many farmers, the choice of *what* to produce may be "factor specific," limited by geography, climate, and soil type. A Kansas wheat grower will never be able to grow pineapples or kiwi fruit, no matter how much more profitable those crops might be. So, for most farmers, the only question is *how much* to produce. Farmer and scholar Fred Kirschenmann sums up the dilemma:

> Rational farmers know that when the price of corn goes down, producing less corn to drive prices up is not a real option. They know that their individual decisions to reduce corn acres in an effort to balance supply with demand will have little effect on supply or price. It will simply reduce their own income. When the price of corn drops, they will produce as much as possible as their only defense against economic disaster. Naturally if the price of corn goes up, they will also produce as much as possible to make up for the income lost in leaner times.[24]

Everyone wants to maximize production, but unless consumption soaks up all that output, the inevitable aggregate result is overproduction—a surplus. Take, for example, this description of a summer 2015 milk glut:

> There's so much milk flowing out of U.S. cows these days that some is ending up in dirt pits because dairies can't find buyers. Domestic output is set to be the highest ever for a fifth straight year. Farmers are still making money as prices tumble because of cheaper and more abundant feed for their herds. Supplies of raw milk are topping capacity at processing plants in parts of the U.S. and compounding a global surplus even with demand improving.[25]

Continued surpluses, if not offset by lower production costs, will eventually *deflate* prices, driving the more financially exposed producers out of business and causing others to reduce production to avoid being overextended. If too many cut back, shortages can result. And the perverse cycle continues. "Neither government bureaucrats nor middlemen have created a vise to squeeze the farmers," wrote agricultural economist Earl Heady in 1967. "The vise is provided by two hallmarks of economic development: on the one hand, by the farmer who produces so much; and on the other hand, by the consumer whose food demand is so inelastic."[26]

Left on their own, then, agricultural commodity markets can be unstable. Even small changes in supply can lead to large swings in prices, as commodity buyers rush to lock in future supplies. As a result, food (unlike smart phones) is seen as too essential to leave entirely to market forces. Most governments seek to manage their agricultural sectors to avoid shortages *and* surpluses and to avoid destabilizing swings in commodity prices. In broad terms, nations want to ensure that their populations have access to sufficient and affordable food. Above all else, they want to avoid sudden and severe shortages, unacceptably high commodity prices, and angry consumers—if not worse. Indeed, in parts of the Middle East, the "Arab Spring" of 2011 may have been sparked less by mass public demands for greater democracy than by fury over dramatic increases in the cost of bread following sudden spikes in global wheat prices.[27] Even in the United States, federal policy since the early 1970s, embodied in successive versions of the Farm Bill, has been designed to ensure that Americans always have access to plentiful, inexpensive food.

Governments also want to avoid surpluses, because unless the excess is somehow absorbed by higher consumption, expanded foreign exports, or new uses (such as converting corn into ethanol), these can push commodity prices below the costs of production, which in turn can destabilize domestic farm sectors and cause economic, social, and political dislocations. A long continuation of the aforementioned milk glut would likely force smaller, less well resourced dairy farms to go out of business, resulting in localized pain. Elected officials don't want to see their constituents hurt, so they agitate for government help.

Finally, no nation wants to let its agricultural sector wither, whether to avoid being wholly dependent on imported food or for fear of losing a connection to an agrarian past that has a defining role in the nation's cultural narrative. Think for a moment about how much of the American (or French or Japanese or Argentinian) story is rooted in an agrarian ideal of self-sufficient farmers and ranchers, and you get the point.[28]

In short, despite professing the ideals of free markets and individual self-sufficiency, most nations manage agriculture, with government policies playing critical roles in influencing what is grown, how much, and at what price. For governments, agriculture—food—is simply too important to be left only to farmers. In the United States, that's where the Farm Bill comes in.

What's in Your Farm Bill? A Brief Overview

The Farm Bill, simply put, is the statutory vehicle through which the US government seeks to manage the nation's food supply and ensure that Americans have abundant and affordable food. Not surprisingly, the Farm Bill affects us all, directly and indirectly, and how it is put together, and by whom, is all about politics. That's why we're here. However, before we get to *why* US agricultural policy looks the way it does—a topic covered in chapter 3—let's take a look at what's in the Farm Bill—or at least the 2008 version of it, which is what Congress had to work with as it approached reauthorization in 2011.

Like a lot of omnibus legislation—a term of art that might be defined as "everything that pertains at least marginally to the policy area under consideration"—the Farm Bill is a sprawling, nearly incomprehensible mess, testimony to a decades-long agglomeration of laws, programs, and rules gathered under the umbrella of the initial authorizing (or "permanent") law each time it came up for renewal. A quick look at the Food, Conservation, and Energy Act of 2008 reveals a lengthy list of distinct "titles," each pertaining to a specific policy area and, as discussed in chapter 4, a distinct set of organized constituent interests seeking to obtain or maintain its particular benefits.[29] Note that many of these titles include laws that were enacted separately and over time were brought under the Farm Bill umbrella to lock in some group's support.

Title I. Commodity Programs. This is the core of the Farm Bill, covering the range of programs and rules designed to support and manage the production of commodities such as corn, wheat, rice, cotton, sugar, and dairy products. In the 2008 bill, such programs included direct payments to farmers to offset the difference between average production costs and market prices, subsidized crop insurance, marketing assistance loans, and other mechanisms to organize specific commodity markets and to manage the problem of imperfect elasticity of demand. Some of these programs originated in the Agricultural Adjustment Act of 1933, arguably the original Farm Bill.

Title II. Conservation. This title includes conservation and wetlands reserve programs (some originating in the 1933 law), many of which were implemented to manage crop production as well as for environmental reasons. For example, the Environmental Quality Incentives Program, devised under the Soil and Water Resources Conservation Act of 1977, offers incentives to farmers to keep ecologically sensitive land out of production, thereby reducing potential surpluses; it also attempts to remediate the effects of industrial-scale agriculture, such as paying for containment basins at CAFOs to keep animal waste from leaching into groundwater.

Title III. Trade. This title covers programs to promote the exportation of US agricultural goods and to fund international food assistance programs, which overall serve to use up commodity surpluses and keep domestic prices stable. Food aid, in this sense, is as much (if not more) about supporting US agriculture as it is about addressing food insecurity in other countries. Programs contained in this title include the Commodity Credit Corporation Act of 1933, the Food for Peace Act of 1954, the Agricultural Trade Act of 1978, and the Food for Progress Act of 1985.[30]

Title IV. Nutrition. Included here is an array of federal nutrition programs, notably the Supplemental Nutrition Assistance Program (SNAP), which can trace its origins to the Food Stamp Act of 1964 and to pilot programs of the late 1930s; the Temporary Emergency Food Assistance Program (1983), which supports community food banks; the commodity supplemental food program; and school lunch programs. Most of these programs were originally designed to use up surplus commodities as much as to feed hungry people.

Title V. Credit. Just as consumers rely on loans or credit cards for large purchases or when cash is tight, a healthy agricultural sector depends on farmers having access to credit to purchase seeds, machinery, and other inputs and repaying those loans once they sell their products. The programs under Title V provide government-backed financing for agricultural production when private-sector credit is unavailable or too expensive, including conservation loans and loan guarantees, the Farm Credit System Insurance Corporation, and emergency loans.

Title VI. Rural Development. This title covers programs designed to extend services to population-thin rural communities, starting with the Rural Electrification Act of 1936 and including the Farm and Rural Development Act of 1961, rural collaborative investment programs, rural health care services, rural housing assistance, Internet access, and grants for water, waste disposal, and wastewater facilities. To a surprising extent, the USDA is as involved in housing and infrastructure development in rural America as the Department of Housing and Urban Affairs is in the nation's cities.

Title VII. Research, Extension, and Related Matters. Title VII programs provide financing for agricultural research and technical training, going back to the Hatch Act of 1887, which established agricultural research stations, and including the National Agricultural Research, Extension, and Teaching Policy Act of 1977; the Food, Agriculture, Conservation, and Trade Act of 1990; and the Agricultural Research, Extension, and Education Reform Act of 1998. As one might expect, this title is of tremendous importance to the nation's vast network of agricultural researchers, most of whom are located at major land-grant universities like Iowa State and Kansas State—themselves created under the auspices of the Morrill Land-Grant College Act of 1862.

Title VIII. Forestry. This title includes the Cooperative Forestry Assistance Act of 1978, which covers the US Forest Service and seeks to foster a healthy forest products sector. Trees are crops too.

Title IX. Energy. This title includes incentives for biofuels production (including corn-based ethanol), as well as funding for research on energy production and conservation.

Title X. Horticulture and Organic Agriculture. Included here are programs for research, plant pest and disease management and disaster prevention, the National Organic Program, and the National Honey Board, created in 1986 to support the production and sale of that sweetener.

Title XI. Crop Insurance and Disaster Assistance. This is an increasingly important title that includes commodity futures regulation (to avoid destabilizing fluctuations in commodity prices), the Small Business Disaster Loan Program, loans and grants to respond to weather and other natural disasters, and other provisions designed to manage the risk inherent in agricultural production.

Title XII. Miscellaneous. This title includes "specialty crops" like maple syrup, programs to support socially disadvantaged and limited-resource producers, biosecurity, livestock health, timber management, and additional provisions on energy and trade. In short, Title XII covers "everything else" that members of Congress deem necessary for policy or political reasons.

Overall, the 663 pages of the Food, Conservation, and Energy Act of 2008 cover just about every facet of agricultural production, processing, transportation, trade, and nutrition that one could imagine. It reflects how the US government—and, by extension, "We the People" through our elected members of Congress—seeks to ensure a healthy agricultural sector and sufficient and affordable food for all. It is also a classic example of omnibus legislation, designed to encompass as many pieces as necessary to ensure its final passage. How the Farm Bill got this way is the focus of the next chapter.

3

History Is Not Bunk

How Farm Bills Past Shape Farm Bills Present

If you took a poll of my farmers and ranchers who have been coming to town meetings for the last two or three years, they would tell you if they had their druthers, they would take the '08 farm bill, scratch out "8" and write "12," but unfortunately, we don't have the money to do that. We're in a dramatically different environment, and that's the world that Collin Peterson and I find ourselves in right now.

—Frank Lucas (R-OK)

Here We Go Again

On August 25, 2011, two members of the Senate Committee on Agriculture, Nutrition, and Forestry found themselves in a hotel ballroom at Wichita's Dwight D. Eisenhower National Airport, convening the second of two "field hearings" on reauthorizing the Food, Conservation, and Energy Act of 2008.[1] Presiding over the half-day session was Debbie Stabenow, a Michigan Democrat in her first year as committee chair. However, the main attraction for those in attendance was the committee's ranking minority member, Kansas Republican Pat Roberts, who had served on the committee since first being elected to the Senate in 1996. Before then, Roberts had spent sixteen years in the House representing the state's First Congressional District, during which time he served on the House Committee on Agriculture and, as its chair, guided House passage of the Farm Bill in 1996.

The presence of Kansas's senior senator, a seasoned veteran of farm policy, assured a good turnout by the state's political, educational, and agricultural establishment. This included Governor Sam Brownback, a former US senator and Kansas secretary of agriculture; the president of Kansas State University, the state's land-grant agricultural college; representatives of various state agricultural organizations, including the Kansas Farm Bureau, Kansas Sunflower Commission, Kansas Livestock Association, Kansas Corn Growers, and Kansas Association of Wheat Growers; and speakers from selected county-level farm credit, rural electrification, and grain cooperatives. Also present, but not formally participating, was Representative Tim Huelskamp, a member of the House Agriculture Committee elected the previous November from Roberts's old district—the "Big First."

The Wichita field hearing was classic congressional theater—or "advertising," in scholar David Mayhew's depiction of this legislative activity[2]—held largely to underscore to home-state voters that Roberts and the committee represented the needs of Kansas agriculture in the nation's capital. Indeed, Stabenow had attracted an equally representative turnout at the committee's other field hearing in May at Michigan State University, the nation's first land-grant agricultural college and, not coincidentally, her alma mater.[3] Roberts, a proud graduate of Kansas State, made certain to promote *his* school, which, "like many of our Nation's land grant institutions, is vital to [the] development and well-being of America's agriculture sector."[4]

On most dimensions, the two senators could not have been more different. Stabenow, first elected to the Senate in 2006 after four years in the House, had a generally liberal voting record on social issues such as same-sex marriage; she backed government actions on economic and environmental problems, so long as they didn't harm her state's auto industry, and in 2010 she supported fellow Democrat Barack Obama in passing his landmark health care reform law: the Patient Protection and Affordable Care Act (ACA). Roberts, by contrast, was conservative on social issues; he tended to advocate for minimal government regulation on economic and environmental matters, and he joined fellow Republicans in opposing passage of the ACA. Yet the two were equally ardent in defending their states' farm sectors and the role of their committee in promoting agriculture overall. Although Michigan is known for manufacturing, it is also a major producer of dairy products, fruits, beans, and sugar beets—the primary source of table sugar in the United States.[5] Kansas is American agriculture writ large—the nation's leading producer of wheat and grain sorghum and a major producer of beef.[6] And, in line with Mayhew's argument that *all* legislative activity has electoral consequences, both were no doubt thinking about their next reelection campaigns—Stabenow in 2012 and Roberts in 2014. As a consequence, both made it clear to the audience in Wichita that they were eager to reauthorize the array of agricultural programs contained in the 2008 bill. Said Roberts, in opening the hearing:

Agriculture faces a tough challenge ahead. . . . We are going to exceed 9 billion people on this planet in the next several decades. . . . In order to meet this demand, agriculture must double our production. Some folks question the need for a Farm Bill with commodity prices where they are today. I do not have to tell this crowd that prices can fall much more quickly than they rise. . . . Without an adequate safety net, plenty of producers struggle to secure operating loans, lines of credit, cover input and equipment costs. We need those producers to stay in business if we are going to meet this global challenge and do so in a way that protects our most valuable resource, our future generations.[7]

Stabenow echoed the challenges and expressed confidence in her committee's ability to balance competing priorities:

This is not going to be easy, but I have a great partner in Senator Roberts and we have a great seasoned Committee, as he and I both emphasize to folks, with more former Chairs of Agriculture Committees and Agriculture Secretaries and Governors and the Chair of the Budget Committee and the Chair of the Finance Committee, and I think if there ever was a time when we had an experienced group of folks in the Senate to be able to focus on the right kind of agricultural policy, I think it is now.[8]

Stabenow's confidence was not misplaced. As table 3.1 illustrates, sixteen of the twenty-one members of Senate Agriculture Committee in 2011 had been in the Senate when Congress passed the Farm Bill in 2008. Eleven had worked on at least two Farm Bills, and nine had worked on three or more, including several who had chaired the committee at one time or another over the past thirty years. In the Senate, consideration of the Farm Bill was in experienced and supportive hands.

Conditions were far more unsettled in the House of Representatives. For one thing, House Democrats had lost sixty-three seats in the November 2010 elections, a proverbial midterm beating for the president's party following

Table 3.1 Members of the Senate Committee on Agriculture, Nutrition, and Forestry, 112th Congress (2011–2012)

	State	Year First Elected
Democrats (11)		
Debbie Stabenow (chair)	MI	2000
Patrick Leahy	VT	1974
Tom Harkin	IA	1984
Kent Conrad	ND	1986
Max Baucus	MT	1978
Ben Nelson	NB	2000
Sharrod Brown	OH	2006
Robert Casey Jr.	PA	2006
Amy Klobuchar	MN	2006
Michael Bennet	CO	2009
Kirsten Gillibrand	NY	2008
Republicans (10)		
Pat Roberts (ranking minority)	KS	1996
Richard Lugar	IN	1976
Thad Cochran	MS	1978
Mitch McConnell	KY	1984
Saxby Chambliss	GA	2002
Mike Johanns	NE	2008
John Boozeman	AR	2010
Charles Grassley	IA	1980
John Thune	SD	2004
John Hoeven	ND	2010

Note: Members are listed in order of seniority on the committee.

the contentious passage of the ACA. Arguably, this reflected the "passions of the moment" the framers of the Constitution had foreseen with the lower chamber.[9] More important, the results shifted control of the House to the Republicans, with John Boehner of Ohio, first elected to the House in 1991, replacing California Democrat Nancy Pelosi as Speaker. In the House Committee on Agriculture, Democrat Collin Peterson of Minnesota's Seventh District, who had shepherded the 2008 bill through the chamber, became the panel's ranking minority member as Republican Frank Lucas, a cattle rancher representing the Third District of Oklahoma, took up the chair. Despite being from different parties, Lucas and Peterson were both social conservatives who shared personal histories rooted in farming, they both represented districts with strong agricultural sectors (sugar beets, wheat, and poultry for Peterson; cattle, corn, and cotton for Lucas), and both had tenures on the House Agriculture Committee going back to the early 1990s. But as Congress convened on January 3, 2011, they encountered a legislative landscape where only eight of the committee's twenty-five Republicans and thirteen of its nineteen Democrats had been in the House in 2008 (see table 3.2). In fact, fifteen of the committee's Republicans were brand new to Congress. Only three of the committee's new majority had experience with more than one Farm Bill. Lucas and Peterson would have their work cut out for them.

The Food, Conservation, and Energy Act of 2008 was eventually approved by overwhelming bipartisan majorities over successive vetoes by President Bush. However, it had been a hard fight that revealed real ideological, regional, and crop-based divisions over the nature, scope, and costs of its various farm and nutrition programs. Many of the law's key provisions would expire as of September 30, 2012, and observers wondered whether renewing it during the 112th Congress was even possible given the already apparent partisan and ideological divisions between a new Republican House majority and a Senate still led by Democrats, not to mention a Democratic president facing what promised to be a tough reelection campaign in 2012. But Stabenow, Roberts, Lucas, and Peterson had no choice: without timely reauthorization, programs important to their constituents would expire a month before Election Day. Like it or not, the Farm Bill was back on the agenda.

Public Problems and Policy Agendas

How do problems get onto the public agenda? At first, such a question might seem odd. Problems become apparent, they are recognized as needing to be fixed, and when it seems clear that solutions require government action, they are dealt with by policymakers. Right? In fact, it isn't that simple. For one thing, with the exception of some great tragedy—the attacks of September 11, 2001, or the devastation wrought by Hurricane Katrina in 2005 come to mind—it usually isn't clear why certain problems, among the many we *could* deal with, make it onto the agenda of government attention and possibly action. After all, there are countless problems—conditions deemed by some-

Table 3.2 Members of the House Committee on Agriculture, 112th Congress (2011–2012)

Republicans (25)	District	Year First Elected	Democrats (19)	District	Year First Elected
Frank D. Lucas (chair)	OK-3	1994	Collin C. Peterson (ranking minority)	MN-7	1990
Bob Goodlatte	VA-6	1992	Tim Holden	PA-17	2003
Tim Johnson	IL-15	2001	Mike McIntyre	NC-7	1996
Steve King	IA-4	2002	Leonard Boswell	IA-3	1997
Randy Neugebauer	TX-19	2003	Joe Baca	CA-43	2003
K. Michael Conaway	TX-11	2004	Dennis Cardoza	CA-18	2003
Jeff Fortenberry	NE-1	2005	David Scott	GA-13	2002
Jean Schmidt	OH-2	2005	Henry R. Cuellar	TX-28	2005
Glenn Thompson	PA-5	2008	Jim Costa	CA-16	2004
Thomas Rooney	FL-16	2008	Timothy J. Walz	MN-1	2006
Marlin Stutzman	IN-3	2010	Kurt Schrader	OR-5	2008
Bob Gibbs	OH-18	2010	Larry Kissell	NC-8	2008
Austin Scott	GA-8	2010	William Owens	NY-23	2008
Scott Tipton	CO-3	2010	Chellie Pingree	ME-1	2008
Steve Southerland	FL-2	2010	Joe Courtney	CT-2	2006
Eric Crawford	AR-1	2010	Peter Welch	VT-at large	2006
Martha Roby	AL-2	2010	Marcia Fudge	OH-11	2008
Tim Huelskamp	KS-1	2010	Terri Sewell	AL-7	2010
Scott Desjarlais	TN-4	2010	Jim McGovern	MA-2	1996
Renee Ellmers	NC-2	2010			
Chris Gibson	NY-19	2010			
Vicky Hartzler	MO-4	2010			
Robert Schilling	IL-17	2010			
Reid Ribble	WI-8	2010			
Kristi Noem	SD-at large	2010			

Note: Members are listed in order of seniority on the committee. The roster does not include Democrat Gregorio Sablan, the delegate from US Micronesia, who had voting rights on the committee but not on the House floor.

one to merit attention—but only so much time, energy, money, or caring to go around.[10]

So why do some problems get our attention but others do not? Political scientist John Kingdon portrays policymaking as a messy "primordial soup" of problems and possible solutions, all competing for policymakers' limited attention and energy. Except in instances of clear crisis, in Kingdon's depiction, most problems get on the agenda only when a "window of opportunity" opens, and this is made possible by the convergence of three otherwise independent "streams": widespread recognition of a problem, ready availability of feasible solutions, and the presence of key policy "entrepreneurs" who are willing to promote policy change and able to mobilize formal approval.[11] Absent these conditions, policy change is less likely.

How might such a dynamic play out? Think for a moment about health care, which is a "problem" for most Americans. However, what *kind* of problem health care poses varies from person to person. For some there is no problem; they are healthy enough or wealthy enough to have no worries about access to adequate and affordable treatment. At the other end of the spectrum are those who lack access to health care altogether. In between are most Americans, who worry about affordable health care as they age, as treatment options become more expensive, or when they change jobs.

But the "problem" of health care has been with us for what seems like forever, so why did Congress take action in 2009 that resulted in passage of the Affordable Care Act of 2010?[12] The short answer is that many more Americans were worried about access to and the costs of health care as a consequence of the economic recession that hit in 2008, the worst since the Great Depression of the 1930s. Feasible ideas to address access and affordability existed, so candidate Barack Obama made health care a key part of his 2008 presidential campaign, and once elected, he had just enough of a Democratic majority in both chambers of Congress to pass his reform package over what became unified Republican opposition. A window of opportunity opened, health care got on the agenda, and major policy change occurred.

Fair enough, but that doesn't quite explain the *substance* of the ACA. Democrats controlled both the White House and Congress when Obama took office, so you might think the ACA was devised by liberal health policy experts in a left-leaning Washington, DC, policy institute. You would be wrong. The individual mandate at the heart of the ACA—which requires

Americans to purchase *private* health insurance—was first proposed in the 1980s by economists at the Heritage Foundation, a "free-market" policy institute, and the idea was first implemented in 2006 in otherwise liberal Massachusetts by Republican governor Mitt Romney, later his party's nominee to oppose President Obama in 2012.

So how did a generally conservative policy idea first put into practice by a Republican governor end up being the centerpiece of a Democratic president's signature health care reform package, even though many of his most ardent supporters expressed a preference for a more "liberal" solution such as Canada's "single-payer" system? The answer is that Obama and Democratic leaders in Congress thought that some version of "Romneycare" was the only feasible option given political and economic realities.[13] It engaged the free market and thus avoided opposition by private health care providers and insurers; it had already been tried by a Republican governor, and it appeared to work; and, by grafting itself onto existing health care systems, it seemed less "radical" or "un-American" than other proposals. In the end, even as Republicans decided to oppose *any* Obama health care plan—calculating (accurately) that doing so would benefit them in the 2010 elections—the proposal was politically palatable enough to get the support of a bare majority of congressional Democrats. So, despite many possible solutions to the "problem" of health care, what became (with some irony) "Obamacare" followed a path of lesser resistance.

The example of the ACA introduces us to the notion of *path dependence*, the reality that policy solutions are rarely imagined out of thin air. Most of the time our policies have long histories, shaped in some way by actions taken years or even decades earlier, and often under very different societal, economic, and political conditions.[14] While such prior actions may not *determine* what we do now—we are always free to start from scratch—the past powerfully affects what we see as feasible. The United States has a history going back to at least the 1940s of providing private health insurance as part of employment benefits, accompanied by the incremental accumulation of public programs to provide care to war veterans through the Veterans Administration, to the elderly through Medicare, and to the poor and disabled through Medicaid. This history imposed boundaries on what the ACA would look like, regardless of some people's dreams for radical change.

US agricultural policy has an equally long path, going back to the Great

Depression. However, unlike health care and other problems in Kingdon's model, agriculture just *shows up* on the congressional to-do list roughly every five years. Why? It can't be because farmers, less than 2 percent of the population, are powerful enough to command routine attention in a country dominated by cities and suburbs. In reality, the regular appearance of agriculture on the legislative agenda is due solely to provisions in the Farm Bill, which requires that Congress formally reauthorize the collection of commodity programs found in Title I roughly every five years, or they will expire on the deadline.[15] Where did those provisions come from? The answer: the Agricultural Act of 1949, amending the Agricultural Adjustment Act of 1938. Why is that, you ask? Time to dive into history.

What follows is a review of US agricultural policy as embodied in successive versions of the Farm Bill going back to the 1930s, taking particular note of the path-dependent nature of current agriculture policy, or how decisions on past versions of the Farm Bill shape current policy debates and impose constraints on the choices available to policymakers. It concludes in early 2011, when Congress again found the Farm Bill on its formal docket—ready or not.

Policy Origins

US agricultural policy today is still defined by the legacy of the Great Depression. The trauma experienced by the nation's farm communities in the late 1920s and early 1930s, depicted in searing photographic images of the Dust Bowl and rural poverty, led to the enactment of the Agricultural Adjustment Acts of 1933 and 1938—the first Farm Bills.

The "problem" of agriculture in the 1930s was, in many ways, the same as it is today: how to avoid price-deflating surpluses while enabling farmers to earn "sufficient" incomes.[16] Prior to that time, US agricultural policy had prioritized *development,* with the federal government seeking to settle the nation's vast interior by providing low-cost or free land to whoever would farm it, fostering agricultural research and general education through land-grant colleges like Michigan State and Kansas State, and providing technical assistance through the USDA's county extension agents, all to ensure the nation's capacity to feed itself.

By most accounts, US farmers enjoyed their greatest relative prosperity in the years just prior to World War I—agriculture's "golden era"—as the demands of the nation's rapidly growing population and robust global export markets meant tight commodity stocks, high commodity prices, and, for most farmers, good incomes. As a result, few farmers wanted government involvement in their activities. However, by the late 1920s, US population growth slowed after the enactment of stringent immigration quotas, and global market demand for exports stagnated even as production continued to climb, leading to expanding surpluses and plummeting commodity prices. Although consumers benefited from lower food costs, successive years of depressed commodity prices and seemingly endless surpluses meant hardship for farmers struggling to bridge the gap between revenue and production costs. As a result, farmers began to call for a more active government role in managing production. Their pleas grew more desperate as the Great Depression devastated rural America.[17]

The Agricultural Adjustment Act of 1933 (AAA), an early part of Franklin Roosevelt's New Deal agenda to tackle the larger economic crisis, for the first time got the federal government involved in *managing* agriculture. Under its rules, farmers agreed to restrict production in return for direct government payments to cover costs and compensate for low commodity prices. Reducing surpluses to keep commodity prices from dropping even lower proved trickier. Early actions to take commodities out of circulation by destroying crops and animals sparked outrage, given that so many Americans were out of work and standing in breadlines. In response, the administration and Congress created the Commodity Credit Corporation (CCC), which purchased and stored surpluses to be sold later when market conditions were more favorable, and the Federal Surplus Relief Corporation (later the Federal Surplus Commodities Corporation), which distributed surplus food to the needy through state and local governments. Such "emergency" use of surpluses to bolster the diets of the poor would form the basis of the school lunch and food stamp programs.

The overarching intent of the AAA, the CCC, and other New Deal actions, including laws restricting the production of sugar, tobacco, and cotton, was to enable the US Department of Agriculture to manage crop production, reduce surpluses, boost commodity prices, maintain farm incomes, and preserve rural communities. Most of all, it was intended to help farmers achieve income parity with what urban Americans earned by offsetting some of the farmers' production costs. However, a special tax on food processors to finance these payments to farmers was declared unconstitutional by the Supreme Court, so in 1936 Congress responded with the Soil Conservation and Domestic Allotment Act, through which the federal government achieved similar results by using general tax revenues to pay farmers to set aside acreage entirely or not to plant "soil-depleting" crops like corn and wheat.[18]

Two years later, Congress gathered these various programs together under the first omnibus Farm Bill, the Agricultural Adjustment Act of 1938, which authorized payments to farmers, set mandatory price supports (or minimum market price levels) for commodities such as wheat and corn, established production quotas and target prices for certain products such as dairy, and implemented a range of production management tools, including acreage allotments, marketing agreements, conservation set-asides, and crop insurance. Importantly, most of these commodity price supports and production quotas were authorized for five years, at which time lawmakers could adjust them to meet changing market conditions.

Post–World War II: The Enduring Problem of Surpluses

The pressure to feed US and Allied soldiers and populations during World War II and to prevent starvation in war-ravaged nations immediately afterward soaked up surpluses and boosted farm incomes. This negated the need for production management or commodity price supports, and although programs put in place by the AAA remained on the books, they weren't necessary. By the late 1940s, however, the problems of surpluses and unstable commodities markets reemerged as other nations rebuilt their agricultural sectors and as American farmers adopted new technologies—synthetic chemical fertilizers and pesticides in particular—to boost production, even as their numbers began to decline.[19]

Congress and the Truman administration responded with the Agricultural Act of 1949, which amended the 1938 act to become, in tandem with it, agriculture's "permanent law," the foundational authority for the commodity programs that still underlie US agricultural policy.[20] "Since then," agricultural journalist Jerry Hagstrom notes, "whenever Congress has passed new farm legislation affecting commodities and dairy products, it has suspended the commodity title of the 1938 and 1949 laws for specific periods of time

Table 3.3 Farm Bills, 1949–2008

1949	Agricultural Act
1954	Agricultural Act
1956	Agricultural Act
1965	Food and Agriculture Act
1970	Agricultural Act
1973	Agriculture and Consumer Protection Act
1977	Food and Agriculture Act
1981	Agriculture and Food Act
1985	Food Security Act
1990	Food, Agriculture, Conservation, and Trade Act
1996	Federal Agricultural Improvement and Reform Act (FAIR)
2002	Farm Security and Rural Reinvestment Act
2008	Food, Conservation, and Energy Act

rather than simply amending a section of the 1938 or 1949 laws or passing new legislation and making it permanent."[21] This seemingly arcane point about suspending provisions of the "permanent" law for a defined period would be critical each time the Farm Bill came up for reauthorization. Table 3.3 lists the Farm Bills from 1949 through 2008.

Continued overproduction and the resulting pressure on smaller farm operations were at the heart of battles between the Eisenhower administration and congressional Democrats through the 1950s. The administration, backed mostly by midwestern Republicans representing corn growers, favored less government intervention and encouraged farmers to "get big or get out." Southern Democrats representing cotton and peanut growers favored the continuation of New Deal–era production controls and subsidy payments. Kansas wheat growers, still reflecting a strain of early-twentieth-century radical populism hostile to corporate power, sided with the Democrats. Equally important, Democrats dominated the congressional committees on agriculture, so farm policy in the 1950s reflected a mix of orientations. Surpluses were absorbed through donations to the needy and to schools and through the Agricultural Trade Development Assistance Act of 1954—also known as the Food for Peace Act—under which the US government purchased surplus commodities and distributed them as international aid. Reauthorization of that act would be rolled into subsequent Farm Bills.

The 1970s and 1980s: Plowing Fencerow to Fencerow

US agricultural policy through the 1960s largely retained its New Deal–era focus on managing production and using up surpluses through domestic nutrition programs, international food aid, and, increasingly, global trade. However, conditions on the farm and in global markets were changing dramatically, leading to major policy changes in the early 1970s.

In 1972 the Nixon administration, seeking to reduce growing stockpiles of surplus commodities and improve relations with its Cold War rival, authorized a massive sale of feed corn and wheat to the Soviet Union. The sale was poorly devised and unexpectedly used up most of the US wheat surplus, just as a drought hit the Grain Belt in the Midwest. That combination of factors led to global grain shortages, sharply higher meat prices, and angry American consumers.[22] In response, newly appointed secretary of agriculture Earl Butz exhorted American farmers to plant "fencerow to fencerow" and leveraged the 1973 bill to shift agricultural policy away from its long emphasis on crop management and toward the maximization of production.[23] To do so, the Agriculture and Consumer Protection Act of 1973 suspended the price supports and management practices established in the permanent law, set guaranteed commodity target prices, and authorized "deficiency" payments to farmers whenever market prices fell below target prices (see table 3.4).[24] Such payments essentially subsidized production, a fundamental policy shift that, as Nadine Lehrer observes, "created an incentive for farmers to sell their crops even if prices were low, since the direct payment would make up their lost income. And it moved farm policy from a system where excess grain was stored when prices were low to a system where it was exported. Since direct payments subsidized farmers without raising the value of the commodities on the world market, farmers could also export these surplus grains more competitively."[25]

Also of note, in putting together the 1973 law, Congress for the first time formally included the federal food stamp program, initially authorized under the Food Stamp Act of 1964; it also expanded program eligibility and mandated that states make benefits available to anyone who qualified. Although eligible recipients still had to purchase their food stamps—each dollar translated into a specified amount in stamps, which could then be used to purchase specific foods—the Farm Bill's new nutrition title extended the

Table 3.4 Types of Commodity Supports

Program Type	Government Action	Effects
Price supports (1930s–early 1970s)	Government buys commodity at a set price to guarantee minimum income for producers; stores surplus until it can be sold when market prices are above target or distributed as aid	Allows high production; sets high commodity prices for farmers and buyers; costs of storing surpluses may be high; guaranteed prices may encourage overproduction
Supply control (1930s–early 1970s)	Government guarantees farmer income by limiting how much can be produced or imported	Limits production; keeps commodity prices high; requires active government management
Deficiency payments (1973–1996)	Government sets a target price and pays producers the difference between it and the market price	Allows unlimited production; supports farm income while keeping costs to buyers low; essentially subsidizes production
Direct payments (1996–2008 Farm Bills)	Government pays producers a set price based on historical production	Subsidizes and guarantees farm income; no direct effect on production volume or prices for buyers
Insurance	Government subsidizes the costs of private insurance or, with *disaster payments*, directly pays producers for losses caused by weather events or other disasters	Essentially guarantees income by enabling producers to avoid losses; no direct effect on production volume or prices for buyers; unexpected events can lead to high expenses
Promotion programs	Government works with commodity producers to promote sales and new uses for commodity	Enables more production; supports producer income; can lead to higher demand but also higher commodity prices for buyers

Source: Adapted from Parke Wilde, *Food Policy in the United States: An Introduction* (London: Routledge, 2013), 22–30, table 2.2.

program to more people than ever and, to satisfy agricultural producers and processors, made more foods eligible for purchase with food stamps. In doing so, Congress formally instituted the "farm programs + food stamps" linkage that would thereafter embody Farm Bill reauthorization. That rural-urban partnership is examined more closely in chapter 4.[26]

From 1973 on, US agricultural policy would emphasize unrestrained production, price supports, and reliance on nutrition programs and export markets to deal with surpluses. It also accelerated the consolidation of farms into larger and more specialized operations, often at significant ecological expense.[27] And, critics argue, it created perverse incentives whereby the USDA, agribusiness, and researchers at the nation's land-grant universities had to work harder to find ways to use up so much output, largely through the introduction of the highly processed "convenience" foods that came to dominate the American diet in the 1980s.[28] Even so, and whatever the later costs in terms of nutrition and the environment, no American president was going to allow commodity shortages to lead to high food prices.

Farm bills over the next two decades, while reflecting momentary shifts in market and political conditions, were adjustments to the 1973 law. In the Food and Agriculture Act of 1977, Congress revived some pre-1973 production management practices in response to farmers' pleas that unabated surpluses were pushing commodity prices far below the costs of production.[29] More important, the 1977 law instituted historic changes in the food stamp program. A bipartisan coalition led by Senator George McGovern, a Democrat from South Dakota who had been his party's 1972 presidential nominee, and Senator Robert Dole, a Republican from Kansas who would be his party's nominee in 1996, leveraged the nutrition title to amend the 1964 Food Stamp Act and eliminate the requirement that needy families purchase their stamps. Food stamps would now be provided at no cost to everyone who was eligible, with the actual allocations based on family income and assets. Also included were new mechanisms for certifying eligibility and preventing fraud. With the purchase requirement removed, food stamp use and program budgets increased dramatically.[30]

Reauthorizations in 1985 and 1990 continued the main thrusts of the 1973 law while responding to growing concerns about the ecological effects of unrestrained production. Conservation set-aside programs were created, and financial incentives were designed to keep farmers from planting on environmentally sensitive lands. That said, Congress left untouched the target prices and direct payments that induced growers to maximize production in the first place.

The 1990s: Freedom to Farm—until Things Go Bad

The next major change in agricultural policy came in 1996 when Republicans, in control of both houses of Congress for the first time in forty years, pushed through the Federal Agricultural Improvement and Reform Act (FAIR). At its core was a so-called freedom-to-farm provision largely promoted by Midwest corn producers and sponsored by Representative Pat Roberts of Kansas, which sought to remove barriers to production and provide incentives for farmers to meet expected increases in global demand.[31] To do so, FAIR again suspended the permanent law's array of New Deal–era commodity price supports, acreage restrictions, and other supply management techniques, this time replacing the 1973 bill's system of annual "deficiency" payments with fixed seven-year contracts to give farmers a predictable income base on which to expand production, regardless of momentary fluctuations in market prices. The ultimate goal, if all went according to plan, was to phase out payments entirely once the contracts ended in 2002. In the meantime, as Fred Kirschenmann notes, FAIR's promoters "assumed that farmers would make adjustments in response to market demand, enabling the government to get out of the subsidy game."[32]

FAIR was a dramatic shift in farm policy. As Lehrer observes, invoking Kingdon's agenda model, its passage was enabled by a convergence of factors:

> First, Republican control of Congress created an atmosphere in which legislators were looking to limit government intervention in agriculture. Second, there was pressure to reduce the growing budget deficit, for example by reducing commodity subsidies. Third, the General Agreement on Trade and Tariffs had highlighted an ideal of liberalized trade to be achieved by countries reducing domestic subsidies and tariffs. Fourth, House Speaker Newt Gingrich (R-GA) authorized commodity programs to be written by budget committees rather than in the more status-quo-oriented agricultural committees. Fifth, the writing of the 1996 bill coincided with a burst of high commodity crop prices in 1995–6.[33]

High commodity prices and healthy profits made farmers less likely to fight to keep deficiency payments, which were already under scrutiny because of their unpredictability, their cost to taxpayers, and their incompatibility with the open global markets envisioned under recent international trade agreements. Equally important, leverage over relevant rulemaking within Congress shifted, momentarily, from promoters of agriculture to fiscal "hawks" on the House and Senate Budget Committees, who were more concerned with attacking the rising costs of commodity programs than satisfying rural constituencies. In fact, House leaders included the freedom-to-farm provisions in a must-pass fiscal-year 1986 budget bill to get around opponents on a deadlocked Agriculture Committee. Congress passed the budget bill in late 1985, only to see it vetoed by President Bill Clinton in part because of his concerns about the costs of fixed payments. Budget negotiations resumed with the commodity programs removed, enabling the Agriculture Committees to regain control and achieve reauthorization through the more or less regular process.

Of interest to our story is that the House version of FAIR, passed in early 1996, sought to repeal outright the permanent law contained in the Agricultural Act of 1949, which Pat Roberts called an anachronistic "fishhook that you had to swallow" if you wanted anything else.[34] Senate Democrats, led by Tom Harkin of Iowa and Kent Conrad of North Dakota, refused to budge; in their view, keeping the permanent law was rural America's guarantee that Congress would revisit farm policy on a regular basis. President Clinton, mindful of agriculture's importance to his home state of Arkansas, agreed. Backed by his veto threat, the House-Senate conference committee followed precedent and suspended the permanent law through 2002.

At first blush, passage of FAIR was achieved through a fundamental reordering of commodity support politics, with consequent impacts on long-standing policy. However, as many agricultural economists warned, expanded production soon outpaced demand, leading to another slump in commodity prices. Producers were soon lobbying Congress to bring back annual deficiency payments to supplement their contract checks. In 2000, with federal budget surpluses lessening concerns about program costs, the Clinton administration resumed deficiency payments on an "emergency" basis, and Congress formally reinstituted them in 2002.[35] Though farmers got their "freedom to farm," it came with a government safety net.

The 2000s: Continuing the Status Quo

The two Farm Bills passed during the George W. Bush presidency reflected legislators' "normal" tendency to incrementally expand programs when given the chance. Reauthorization in 2002 benefited from the relative lack of concern about the budget—the Bush administration took office in 2001 enjoying the first federal budget surplus since the 1960s—and from its low position on the president's list of priorities in the wake of the September 11 attacks. Left largely alone, Agriculture Committee members reinstituted the annual direct payments suspended under FAIR and created new incentives for biofuels to soak up surplus corn stocks and, they argued, to facilitate "energy independence" from Middle Eastern oil. Critics argued that by subsidizing corn ethanol, the Republican-controlled Congress only made bad energy and environmental policy. More broadly, they argued that the bill did little to instill market discipline or address overall farm incomes; two-thirds of all direct payments went to 10 percent of US farm owners—the largest and most well capitalized—and encouraged continued overproduction at the expense of small and medium-sized farms.[36] But President Bush was occupied with foreign policy and made little effort to rein in his own party on what observers regarded as a classic congressional "Christmas tree"—something for everyone.

The context was far different in 2008. Increasing commodity and nutrition program costs, combined with the reemergence of large federal budget deficits in the wake of revenue losses from tax cuts and the costs of expensive wars in Iraq and Afghanistan, prompted a fight between Bush and a Congress that was now controlled by Democrats. However, an unpopular president in his last year in office was ill positioned to influence a bill that was important to legislators in both parties as the fall elections neared. Congress made modest adjustments to commodity programs, instituted insurance and research programs for fruits and vegetables not covered in the permanent law, increased support for biofuels, and, perhaps most important, broadened eligibility for food stamps under the renamed Supplemental Nutrition Assistance Program (SNAP). Bush opposed the bill as too expensive, but members of Congress in both parties banded together and easily overrode his veto. (Actually, they did so *twice*. Clerical errors in the bill as enacted required Congress to pass it a second time. Bush vetoed it again and was again overruled by wide margins.) The Food, Conservation, and Energy Act reauthorized farm and nutrition programs through fiscal year 2012.

Footprints

This review of US agricultural policy reminds us of the core dilemmas facing the House and Senate Committees on Agriculture as they set out to reauthorize the Farm Bill. In doing so, they would be confronted by the legacies of decisions made as far back as 1933.

By 2011, the omnibus bill cost $85 billion annually to cover a sprawling maze of programs enacted in reauthorizations going back to 1949 or included after first being enacted as separate legislation. As noted in chapter 2, the Food, Conservation, and Energy Act of 2008 ran over 660 pages and covered fifteen titles, each numbing in its technical detail and reflecting bargains made (and sometimes forgotten) to gain passage of some previous reauthorization in decades past. Most notable among them was SNAP, which by 2011 had grown to be the single most expensive part of the Farm Bill. As discussed in chapter 4, it had been included in the 1973 bill to broker an exchange: the votes of House members from urban areas to support commodity programs, in return for the votes of rural members to support nutrition assistance for the needy. As such, any discussion of Title IV would confront decades of prior decisions about program eligibility, along with a coalition of support inside and outside Congress for maintaining this essential safety net for the food insecure. Other titles came with their own coalitions of supporters, each ready to defend its respective piece of the Farm Bill pie. In short, those who revisited the Farm Bill in early 2011 were not starting from zero.

Among these legacies, perhaps the most important is one that few outside of agricultural policymaking understand or appreciate. As noted earlier, unlike many federal statutes—such as many of the 1970s-era laws that still define US environmental policy[37]—each Farm Bill is technically a *reauthorization* of the permanent law, the Agricultural Adjustment Act of 1938 as modified by the Agricultural Adjustment Act of 1949. Making commodity programs temporary was intended to foster experimentation, to make it easier to revise problematic subsidy formulas, and to keep agricultural policy current as production and market conditions changed.[38]

So, you might ask, why doesn't Congress just change or amend the "permanent" law rather than "suspend" it for a defined period of time? After all, if Congress fails to reauthorize the bill by the specified date, in theory, many of the commodity price supports and marketing rules first instituted in 1938 or 1949 would go back into effect, regardless of how much agriculture or society has changed in the ensuing decades. "Without reauthorization," Tim Huelskamp observed in his doctoral dissertation, "agricultural policy would revert to New Deal–era legislation, the consequences of which would be an immediate increase in budgetary expenditures, tight and perhaps unconstitutional constraints on farmers' planting and harvesting options, and a substantial escalation in consumer food prices."[39] Crop supports without authorization in the permanent law, such as those for peanuts, and any programs directed at organic crops would expire. To repeat an oft-cited horror story, failure to reauthorize would force the USDA to purchase bulk milk at 1949 prices, causing consumer dairy prices to *triple* overnight. Of course, as Huelskamp hastened to note, Congress and the president could stave off the direst effects of program expiration by other means, including short-term reauthorizations. But such stopgap efforts might not work and, at a minimum, would induce a degree of unwelcome program—and economic—instability.

So why do those who care about agriculture continue to play what is, in effect, a routinely scheduled game of legislative "chicken," daring Congress *not* to reauthorize? Politics. For decades, the prospect of remanding agricultural policy to some New Deal–era "state of nature" was a useful doomsday threat promoted by defenders of agriculture to force everyone to the table and to gain buy-in for new programs and spending. More recently, it seems that the ever-shrinking congressional farm bloc doesn't *dare* put the Farm Bill up for a full revision. With Congress, particularly the House, filled with legislators from urban and suburban constituencies, the relative few who farm are not inclined to entrust their government safety net to the whims of eaters. But, as we will see, this has become an increasingly perilous task.

4

Whatever It Takes

Farmers, Food Stamps, and Coalitions of Convenience

The bill is huge! . . . There isn't anything in American agriculture, farming, and health that this bill doesn't touch, but there is no overarching agenda. The Farm Bill is simply a collection of government-supported programs, each with its own collection of lobbyists, proponents, and opposing forces. You get the sense that everyone said, "Let's just throw this program in." There is nothing rational in the Farm Bill.

—Marion Nestle

Had you sat in on any of the field hearings convened by the House Committee on Agriculture in early 2012, you likely came away thinking that only farmers cared about the Farm Bill. Depending on which hearing you observed, the speakers would have testified to the unique problems facing corn, soybeans, rice, wheat, hay, sorghum, pork, beef, milk, cotton, catfish, peanuts, potatoes, apples—even wine grapes—and why renewing the Farm Bill was essential to their well-being. You might have gotten a slightly broader picture had you attended one of the Senate committee's field hearings, which in addition to producers included speakers addressing the need for Congress to continue funding for agricultural research, farm credit, rural economic development, food safety, conservation, and, in one instance, nutrition, even as elsewhere congressional leaders wrangled over making major cuts in overall federal spending.

In some ways, the contrasts between the respective field hearings (table 4.1) reflected the constitutionally imposed differences be-

Table 4.1 Testimony at House and Senate Field Hearings, 2011–2012

	House		Senate		All Hearings	
Speaker Type*	Number	Percent	Number	Percent	Number	Percent
Local producers	38	97.4	8	22.2	46	61.3
Farm and commodity groups	1	2.6	14	38.9	15	20.0
Local businesses and institutions	0	0	7	19.4	7	9.3
Conservation	0	0	3	8.3	3	4.0
Research and policy	0	0	2	5.6	2	2.7
Rural development	0	0	1	2.7	1	1.3
Anti–hunger/nutrition	0	0	1	2.7	1	1.3

*Some speakers represented two types of organizations and were counted twice.

tween the chambers. The House, whose members represent districts of equal population size for two-year terms, is, by design, more local and more focused on the immediate, compared with the arguably more expansive views of senators serving statewide constituencies for six years. Even so, the picture from either roster reflected what one would imagine agriculture to be: farmers, and affected local businesses and institutions.

But that picture was nowhere near representative. Of note was a longer, more diverse list of those who cared about the Farm Bill but were *not* invited to speak. Even as it convened its hearings, the House Committee on Agriculture hosted an online portal through which anyone could comment about the Farm Bill or about agriculture and food policy overall.[1] Some 5,000 individuals took the time to send in comments, which, if one takes (a lot of) time to read, provide a fascinating and very different take on what some Americans thought Congress should do about the food system. Four examples, listed in order on page 522 of the transcript, are illustrative:[2]

COMMENT OF GAIL CONLEY
Date Submitted: Sunday, May 20, 2012, 6:06 a.m.
City, State: Fairfax, VA
Occupation: Retired
Comment: This nation was built on the backs of the small farmers. Don't turn your backs on them at a time like this when they need your help. We are a nation in need, our people are obese and getting sicker as a group. Those who seek

to help them with healthful foods are being squelched. It is [in] your hands to do the right thing for the future generations of organic farming. Give us all a chance, please!

COMMENT OF HILARY CONNAUGHTON
Date Submitted: Saturday, May 19, 2012, 10:24 p.m.
City, State: McCloud, CA
Occupation: Nurse
Comment: I'm tired of feeling like our agriculture is just another greedy corporation and I want government that I feel wants what is best for me and mine.

COMMENT OF CASEY CONNELL
Date Submitted: Friday, May 11, 2012, 12:49 p.m.
City, State: Belen, NM
Producer/Nonproducer: Producer
Type: Dry beans & peas, vegetables
Size: Less than 50 acres
Comment: This country needs healthier food. We need to practice sustainable agriculture so that our future generations will have a planet they can continue to use for food. One that has not been poisoned by pesticides and herbicides which contains the vitamins and minerals we need to be healthy.

COMMENT OF JOANNA CONRARDY
Date Submitted: Wednesday, May 2, 2012, 9:34 p.m.
City, State: Albuquerque, NM
Occupation: Educational assistant
Comment:

1. Fairness for small farmers is essential.
2. Labeling Genetically Modified food products is essential.
3. Helping the poor and nearly poor adults and children in this country is essential. Please act responsibly when voting on the farm bill. Thank you.

So other voices *were* out there—many more than one might guess by observing only the field hearings.

Another example: On Sunday, January 29, 2012, some 350 Boston-area residents—yours truly included—jammed an auditorium at the Museum of Science for a three-hour "teach-in" on the Farm Bill.[3] Few of those present that winter afternoon grew more than vegetables in their backyard gardens, but they all came because of concerns about the food system and interest in the event's featured speakers. One was Representative Chellie Pingree, a

Democrat from Maine's First District, a member of the House Committee on Agriculture, and a former organic farmer on an island just off the Maine coast. In terms that were unlikely to resonate with her midwestern Corn Belt colleagues, Pingree argued for shifting the Farm Bill's traditional emphasis on commodity crops to helping smaller farms typical of New England and addressing concerns about the food system expressed by consumers—the eaters in the audience. So skewed was federal policy, Pingree told her listeners, that the 2008 bill had been the first *ever* to consider of the needs of "specialty crop" growers like herself. Even so, she added, that bill barely made a dent in the direct payments going to commodity crops, which was why she had just introduced her own Local Farms, Food, and Jobs Act (H.R. 3286) for consideration in the months to come.

As a member of Congress with a direct role in Farm Bill reauthorization, Pingree was an important speaker, but the real draw for most attendees was someone unlikely to show up on an Agriculture Committee hearing roster: Marion Nestle, professor of nutrition, food studies, and public health at New York University and an unsparing critic of the food system and its promoters inside and outside government. Nestle was there to educate her urban audience on what was contained in the current Farm Bill, how it affected the types and costs of food produced—for example, how programs promoting the production of corn essentially subsidize the production of inexpensive sugar—and with what broader economic, environmental, and public health effects.[4] Given such effects, Nestle exhorted attendees not to leave Farm Bill renewal up to traditional agricultural interests, to contact their members of Congress, contribute to groups advocating for food system reform, speak up to shield nutrition programs like SNAP from budget cuts, and spread the word to fellow eaters.

The implicit theme of the Boston teach-in might well have been, "the Farm Bill is too important to leave to farmers." Few in the room blamed individual growers for the state of the nation's food system—and certainly not the small family farms dotting the New England landscape—but they agreed that the Farm Bill should no longer be the preserve of Big Corn, Big Cotton, Big Sugar, and Big Beef. Eaters wanted to have their say.

In 1933, 1949, 1965, and even 1973 it was possible to convene a few men—and they were mostly men—in a Capitol Hill committee room and draft a new Farm Bill.[5] This was no longer true by 2012, as evidenced by the thousands of comments sent to the House Agriculture Committee and the hundreds who attended the Boston teach-in. All over the nation—and in other nations as well—a strikingly large and diverse array of individuals, organizations, and institutions was poring over details of the 2008 law, spreading the word to allies, and getting into position to promote—or prevent—changes in 2012. Recall how the titles of the various Farm Bills have shifted over time from simply the "Agricultural Act" to, by the 1970s, something different each time the Farm Bill was reauthorized. Those titles weren't merely symbolic: they shed light on the issues and related organized interests that had to be considered, worried about, and ultimately included at the bargaining table by Farm Bill promoters to guarantee passage.

In this chapter I trace the expansion of the spectrum of organized groups seeking to stake a claim in some part of the Farm Bill. I start by looking at the origins of the "farm lobby" and how that "cozy triangle" of commodity producer groups, federal bureaucrats, and members of the Agriculture Committees maintained a tight hold on farm policy for decades. Their grasp loosened after World War II as Americans headed to the cities and suburbs, a demographic shift that reshaped the nation's political landscape and impacted representation in Congress. As a consequence, by the 1960s, a shrinking farm bloc needed allies in urban America to maintain its leverage on agricultural policy, and it found them through a most unlikely vehicle: food stamps.

By the 1980s, still others were demanding to have their say as more Americans who did not farm began to pay attention to the health, environmental, and economic impacts of food production. Each new wave of claimants posed new challenges to the food policy status quo, forcing the farm bloc to work that much harder to make whatever deals were necessary to gather enough votes to get a bill to the president's desk. Farm Bill titles are suggestive of those deals, which became part of an ever-expanding "omnibus" law. Whatever it took.

By early 2012, as the Agriculture Committees readied themselves to reauthorize the 2008 law, activists and organizations claiming to speak for virtually every facet of society were positioning themselves for the dealmaking to come: environmentalists, nutritionists, food retailers, animal welfare advocates, fast-food chains, commodity brokers, rural credit unions, university researchers, international trade experts, hunger relief groups, alternative energy promoters—even farmers. Each of these organized interests, often representing opposing sides on the question of whether to continue one program or another, had their supporters in Congress. As a result, everyone knew that stitching together a coalition big enough to pass a bill was going to take more than a few men meeting in a committee room.

The Farm Lobby

As the simple titles of the first several Farm Bills suggest, in those early years, the groups in society that cared about agricultural policy were those one might expect: crop producers and affected local businesses and institutions, from community banks and farm equipment sellers to chambers of commerce and school boards. This "farm bloc" was represented in the federal government by the US Department of Agriculture, locally through Agricultural Extension Service agents, and in Washington by USDA officials promoting farmers in the bureaucratic byways of the nation's capital. As Wesley McCune notes:

> More than any other Cabinet department, Agriculture is the protagonist, the pleader for its constituents. It teaches them to be self-sufficient, to use government credit, and to live inexpensively. Like the Department of Labor and the Department of Commerce, it grinds its ax officially for one part of the national economy—only more so. That is not to say that it should not do so; it is only to say that *the farm bloc is not made of thin air*. It is nurtured, consciously or unconsciously, twenty-four hours a day. It is a facet of the democratic way of lending a helping hand to one chink of society.[6]

The extent to which agriculture depends on its USDA patron is not appreciated in urban America, where for most the federal government is a distant creature. Not so in the Farm Belt, where federal policy has direct consequences—so direct, in fact, that the national organizations advocating for agriculture did not emerge "spontaneously" out of some grassroots activism. They instead were the offspring of federal policy, in some instances organized outright by USDA bureaucrats. The American Farm Bureau Federation (AFBF), the largest general-purpose farmers' group, came into being after passage of the 1914 Smith-Lever Act establishing the Agricultural Extension Service. The Extension Service relied on a network of county agents to bring modern science and technology to the farm and, with the USDA's blessing, to organize notoriously individualistic farmers to act on their own collective behalf.[7] They did so through the development of a parallel network of county- and state-level "farm bureaus," which by the 1940s represented 1.2 million farm families and were arguably the most important secular organizations in rural America.[8]

The AFBF, or Farm Bureau, was organized in 1920, and its national breadth and links to local bureaus made it a potent force in American agriculture through the 1950s, particularly in the politically important midwestern Grain Belt. From its origins, its critics argued that the Farm Bureau had been "conceived by businessmen and county agents, born in a chamber of commerce, nurtured on funds from industry, and has never completely left its home and its parents."[9] Such criticism stemmed from the bureau's unromantic view of farmers as producers for the market, which led it to support large-scale, technologized monoculture. As David Danbom observes in his study of early-twentieth-century American agriculture, "the Farm Bureau embraced the same ends for agriculture in which the USDA believed."[10]

Tied to these farm bureaus were farmers' cooperatives, collective enterprises that helped members market their crops at more competitive prices and got them deals on equipment and fertilizer. By 1940, the National Council of Farmer Cooperatives represented some 4,500 affiliated local co-ops, most of them in the midwestern Wheat and Corn Belts.[11] Whereas the local bureau gave farmers information and technical expertise, the co-op gave them marketing and purchasing power, a role that became increasingly large scale and profitable as agriculture evolved toward ever bigger farms.

From their inception, both organizations were structured as federations, with local and state chapters that included representatives from governments, businesses, and colleges. This structure was effective both in Washington—each national organization speaking to the USDA and to members of Congress on behalf of *all* farmers—and at home, as each county or state chapter connected to its respective representative or senator on more localized concerns.[12] That both organizations mirrored the decentralization and local ties embedded in the American federal system was no coincidence, a reminder that the structure of the political system itself affects how interest groups organize and gain access to policymakers.

Incentives: The Rise of Commodity Groups

If the farm organizations' size and structure served them well in Washington, their popularity with farmers stemmed more from the direct benefits they provided. After all, one did not need to belong to the Farm Bureau or to the National Farmers Union (which tended to represent small farm operations) to benefit from stable commodity prices or supportive federal policies. One need only be a farmer or a resident of a farm community affected by policies made in Washington. So why join?

This is where the notion of incentives comes into play in considering why some "interests" are well organized and well represented in politics but others are not. The task was easier for the farm cooperatives, which, as commercial enterprises, were designed to get their members better prices for their commodities. Becoming a member and paying dues made economic sense. But the Farm Bureau and other general farm organizations, in seeking to represent *all* farmers, had to rely on a range of "particularistic" benefits to maintain member loyalty.[13] AFBF members got special life insurance rates, discounts on merchandise, and other direct benefits to complement the organization's policy objectives. Such incentives also eased the task of mobilizing farmers to take action—such as asking them to contact their members of Congress—because they were better able to identify their direct stakes in the organization's goals.

If farmers could perceive some common "interest" in organizing to influence a particular Farm Bill, what about consumers—the eaters? For them (us?) it's not so easy, and not because they (we?) lack knowledge of or interest in the issues. It's about unequal incentives: any farmer has a far greater direct economic stake in the outcome of the Farm Bill than do most of us consumers, for whom the direct costs of commodity programs might be pennies in extra food expenses or additional federal budget outlays. Even for the most ardent food activist, the personal incentive to take action pales when compared with that of anyone making a living in food production.

Such asymmetrical incentives affect advocacy groups' ability to pull together and mobilize their supporters. In this regard, political scientist E. E. Schattschneider offers a useful distinction between "private" and "public" interests.[14] Private interests such as the Farm Bureau generally promote "exclusive" benefits—higher farm incomes—that, regardless of their portrayal as "good" public policy, accrue mostly to group members or to some defined sector of society. "Public" interests, by contrast, seek "inclusive" benefits that everyone in society can enjoy, regardless of organizational membership or social status.

Using this distinction, producer groups advocating higher prices for feed corn are seeking "private" goods, because any benefit goes almost exclusively to corn growers. Those raising beef cattle, by contrast, want *low* corn prices and have every incentive to oppose policies that keep prices high. Both are "farmers," but they have opposite economic stakes in policy outcomes. Each group is also likely to care more about its particular narrow slice of the Farm Bill pie than about "farm policy" in general—a classic case of "interest group politics" whereby one organized group stands to win at the expense of another (corn versus beef, in this case).[15] As a result, by the mid-1950s, general farm groups like the Farm Bureau were starting to lose the competition for farmer loyalties to narrowly organized commodity groups like the National Corn Growers Association, American Soybean Association, and National Cattlemen's Beef Association, which, by unifying like producers, could promise members clear and direct economic benefits.[16] But even those producing similar commodities can be divided: the large-scale dairy operations represented by the National Milk Producers Federation tend to oppose quotas or price floors on milk production, while small dairy farms that cannot compete with larger producers on a pure price basis tend to belong to the National Dairy Council.

Even so, the problem of providing incentives for any farm group pales in comparison to the difficulty of organizing consumers. Most obviously, not all eaters are alike—for many, food price, quality, and availability are not problems—and it is difficult to find a common theme to mobilize consumers to action beyond vague goals such "keeping food prices low" or "keeping food safe." Such rallying cries might work in a moment of crisis—for example, when pathogens in fast-food beef cause illness and death[17]—but for the most part, it is hard to mobilize the mass of consumers for sustained collective action. More important, "safe food" is a "public good." As such, it is indivisible. If the food is safe, it is safe for *everyone*, even the majority who did nothing to make it safe. So, with the possible exception of someone directly harmed by unsafe food or agriculture-related pollution, most of us have little incentive beyond a spirit of civic-mindedness to become involved in promoting safe food or clean water. Why join a group, go to a teach-in on the Farm Bill, or contact your member of Congress if you benefit anyway?

Economists call this the "free-rider" problem.[18] Even groups representing farmers are subject to it, which is why the Farm Bureau offers members discounted life insurance and other selective benefits. For commodity groups like the National Cattlemen's Beef Association, the solution is to boost overall sales for that commodity by getting supportive members of Congress to insert provisions into the Farm Bill authorizing sector-wide marketing ("check-off") programs subsidized by fees levied on producers. For example, the famous marketing campaigns "Beef: It's What's for Dinner" and "Got Milk?" are backed by USDA-administered check-off programs paid for by fees levied on beef and dairy producers, respectively.[19] Arguably, each individual producer would benefit from higher overall sales but has little incentive to pay for marketing if the benefit can be obtained anyway. Yet the marketing campaign could not occur at all unless *someone* paid for it. Thus, the USDA—at the behest of the industry—levies the mandatory check-off fee to avoid free-riding.

Overcoming the free-rider problem is toughest for organizations promoting public goods such as safe food, the better treatment of animals, or less environmentally destructive farming practices. As Mancur Olson observes in *The Logic of Collective Action*, groups advocating for private goods can mobilize members because the benefits they seek are direct and tangible—higher corn prices equal higher profits—whereas for those advocating for public goods such as clean water, the benefits are often indirect and less tangible.[20] Think about it: humans, not hens, must organize and advocate for better conditions in egg production, even though, for some people, the resulting egg is more expensive and doesn't taste any different.

Incentives explain why groups seeking private goods historically dominate Farm Bill action. It is simply harder to mobilize eaters on behalf of some public good. Such biases in favor of private interests are reinforced in Congress, with different commodity and agricultural groups being represented in particular states and House districts. As such, and even absent campaign donations to friendly legislators, cotton growers will always enjoy easy access to and support from southern members of Congress in districts and states where cotton grows, just as wheat producers will always get the ears of members whose constituencies span the midwestern Wheat Belt. Self-interest—both economic and political—creates common bonds.

Such mutual self-interest manifests itself in telling ways. For example, by the mid-1950s, the House and Senate Agriculture Committees had organized themselves into commodity-oriented subcommittees so that their members could better respond to the distinct needs of corn, cotton, or dairy.[21] Those subcommittees worked closely with commodity groups, which, as John Mark Hansen notes, "had come of age in defending their particular, specialized, and generally non-ideological interests."[22] In contrast, the Farm Bureau tended to reflect the free-market views held by Republicans, and the National Farmers Union tended to side with the production management orientation favored by Democrats. As a result, by the 1960s, Farm Bill politics centered less around the ideological debates of the 1950s, when Democrats defended and Republicans attacked the continuation of New Deal–era production controls, and more around the pragmatic stapling together of deals cut by the subcommittees for their respective commodities, each tacitly agreeing not to get in the other's way: if you don't block support for my cotton, I won't block support for your corn. The full committee, as Douglas Cater observed in 1964, simply "resolves itself into combining the various subcommittees' reports to make an 'omnibus' farm bill, thus frustrating the best efforts of an Administration, Republican or Democratic, to promote a farm program that makes sense in overall terms."[23]

The coziness of these relationships among commodity groups, supportive midlevel USDA officials, and members of the relevant congressional agricultural subcommittees led students of politics to write a series of classic works in the 1950s and 1960s on clientele-based "subgovernments" and "subsystems." Such a configuration, J. Leiper Freeman argued in 1955, "refers to the pattern of interactions of participants, or actors, involved in making decisions in a special area of public policy . . . found in an immediate setting formed by an executive bureau and congressional committees, with special interest groups intimately attached."[24] A decade later, Cater wrote about these "cozy little triangles" of policymaking: "More than mere specialization, the subcommittee system permits development of *tight little cadres* of special interest legislators and gives them great leverage. A conspicuous example, the House Agriculture Committee rarely contains more than a member or two representing the urban consumer. It concentrates primarily on reconciling the various agricultural interests. Democratic members are more disposed toward cotton, tobacco, peanuts, and rice; Republicans lean toward corn and wheat."[25] Cater's observation about the lack of urban representation on the House Agriculture Committee underscores that incentives not only dictate which groups in society seek to be included in Farm Bill negotiations but also influence which members of Congress care enough to pursue seats on the "little legislatures" with direct responsibility over lawmaking.[26] Until the 1960s, most of those members were from the farm bloc, with consequent impacts on food policy.

Farmers and Food Stamps

John F. Kennedy's narrow victory in 1960 came about because of urban and suburban voters, punctuating the shift in the nation's population—and political power—away from rural America. That shift also reshaped the political calculus on agricultural policy: Kennedy saw little political benefit in spending large amounts of taxpayer dollars on commodity programs for a decreasing number of farmers. His sentiments were shared by many legislators from outside the Farm Belt, for whom farmers were just another special-interest group, regardless of their otherwise privileged status in American popular culture.[27] In turn, members of Congress representing farm constituencies worried about their ability to maintain support for the bundle of commodity programs on which the nation's growers and agribusinesses had come to depend.[28]

The new political reality became painfully apparent to the farm bloc in 1964. Liberal Democrats from urban areas were outraged when the rural conservative southern Democrats and midwestern Republicans in control of the House Committee on Agriculture blocked a bill proposed by President Lyndon Johnson to establish a permanent food stamp program, even as those same legislators promoted a measure to increase federal spending on cotton and wheat programs. Johnson, mindful of the growing clout of urban and suburban liberals in the Democratic Party, which at the time controlled both chambers of Congress, called the farm bloc's bluff: report the Food Stamp Act out of committee, or watch the cotton and wheat bill go down to defeat on the House floor. Hansen describes what happened: "Under President Lyndon B. Johnson's not-so-gentle prodding, House Agriculture Committee chairman Harold D. Cooley pried an authorization for a permanent food stamp program out of his committee, which had long been inhospitable to any welfare programs that did not benefit farmers. On the floor, the urban liberals insisted that the food stamp vote precede the farm bill vote, lest unrepentant conservatives defect. The script went as planned."[29]

This explicit "logroll" on two otherwise imperiled bills was a clear lesson for the farm bloc, particularly in the House: no bill supporting commodity programs would ever get enough votes beyond the Agriculture Committees unless it also did something for nutrition.[30] As Randall Ripley notes, "the food stamp program was the one food program that had even a partial urban orientation to it. Urban members skeptical about billions for farm programs could receive at least mild comfort from the prospect of millions for the urban needy."[31] Members of the Agriculture Committees got the message, despite being, as Tim Huelskamp points out, "strongly, ideologically opposed to the food stamp program."[32] Self-interest would trump ideological opposition to "welfare." Everyone had constituents to support and reelection to win.

So clear was this message that, for the first time, the House Committee on Agriculture formally included the Food Stamp Act as a title within the tactically named Agriculture and Consumer Protection Act of 1973, which went on to easy final passage. Very soon, Hansen says, food stamps were *just another commodity*, as far as committee members were concerned, one to be bundled with corn, cotton, dairy, and soy into the omnibus Farm Bill to gain the necessary votes in committee and on the increasingly suburban House floor.[33] Huelskamp quotes one House Agriculture Committee member:

> During my tenure, we made the food stamp bill part of the farm bill. It was a carefully calculated thing which was done a long time ago to try and unite urban interests with agricultural interests, in common support of the bills, that had been fighting each other over. You got a food subsidy program, you got a farm subsidy program. Each in their own are unpopular with the other segment of society, but you put them together and you can get a lot of people who will vote for both of them that way.[34]

A parallel dynamic played out in the Senate Agriculture Committee, which expanded food stamp eligibility in its markup of the 1973 bill. More notably, in 1977 the Senate formally put nutrition programs under the jurisdiction of its newly renamed Committee on Agriculture *and Nutrition*. The informal logroll was now institutionalized.

Linking the fate of the Farm Bill to its treatment of food stamps had clear political and policy benefits, as Marion Nestle observed to her audience in Boston in 2012:

> The elephant in the room in the Farm Bill is food stamps. Forty-five million Americans get food stamps, and it overshadows the amount of money spent on agriculture subsidies by many orders of magnitude. What are they doing in the Farm Bill? Why isn't there a separate food bill? Here's the deal: politicians couldn't get the votes for farm subsidies unless they got votes and support for urban programs.[35]

A shrinking farm bloc would enlist its urban and suburban colleagues in supporting a bill they might otherwise oppose, given the comparative generosity of subsidies going to a small group of farmers. In return, promoters of nutrition programs got the otherwise reluctant support of their conservative rural colleagues. As John Peters summarizes, the food stamp program was a "ready-made issue on which to trade rural votes on food stamps for urban votes on the commodity portions of the bill."[36] Small wonder that for decades thereafter the Farm Bill embodied classic political science notions of legislative logrolling and distributive politics.[37] Everyone got something.

However, logrolling on food stamps wasn't just a particularly colorful instance of vote trading. In linking their fate to passage of the Farm Bill, promoters of federal nutrition programs had gained a unique hole card. Remember, each Farm Bill is technically a reauthorization of the permanent law, requiring Congress to renew its commodity programs within a defined period. Although foods stamps were defined as an entitlement under the Food Stamp Act, and although program funding could be extended separately through the congressional appropriations process, program defenders never saw that route as politically viable given the tenuous social (and often racial) status of food stamp recipients. Instead, nutrition advocates lashed themselves to the mast of a legislative ship that the farm bloc could not allow to sink.

This awkward marriage of convenience would hold up for decades, giving birth to winning coalitions for successive reauthorizations of the Farm Bill in the face of growing changes in agriculture's external economic, demographic, fiscal, and political conditions. However, as Huelskamp observes: "Rather than the limited and closed policymaking networks of earlier times dominated by the traditional farm interests, particularly the commodity organizations and general farm groups, by the late 1970s and throughout the 1980s and into 1990, the external environment has become increasingly diverse and open, complex and unstable, and conflictive."[38]

The sharpest conflicts would revolve around the food stamp program—the awkward marriage of convenience—which by the mid-1990s had already become the Farm Bill's single most expensive title and an increasingly visible target for conservatives looking to cut spending. As a result, "what to do about food stamps" became *the* central problem in Farm Bill reauthorization, one that was typically managed by Agriculture Committee leaders based on their strategic calculus of whether including food stamps helped or hindered the Farm Bill's progress in the larger body.

That tension broke into the open in 1996 when House Republicans, in control for the first time since 1954, sought to segregate food stamps in a separate bill, even as they set a "free-market" direction for farm policy in the Federal Agricultural Improvement and Reform Act (FAIR). Unhitched from commodity programs, their logic went, a food stamp program that no longer enjoyed formal logrolling status within the Farm Bill would be easier to cut back.

Although this was the first time House leaders decoupled food stamps from commodity programs, chairs of the Senate Agriculture and Nutrition Committee had done so earlier, but primarily to manage competing factions within the larger chamber. In 1981 Republican Jesse Helms of North Carolina reported out separate farm and nutrition bills to obtain passage in the Senate, at that time controlled by his party, with the implicit understanding that the two would be reconnected in conference committee to ensure passage in the Democratic-led House.[39] Helms, who otherwise loathed food stamps, restored the program's formal logrolling role in the 1985 bill, when Senate passage was less assured amidst splits within the farm bloc induced by a weak agricultural economy and the added pressures of planned federal budget cuts. Five years later, Patrick Leahy of Vermont, a liberal Democrat, decoupled nutrition and commodity programs in the Senate committee, planning to reconnect them in a floor amendment encompassing a more comprehensive set of nutrition programs. However, he held off on the amendment to avoid tangling with the Gramm-Rudman antideficit law, which mandated no net increase in federal spending, and nutrition programs were reconnected to the Farm Bill in a House-Senate conference.

In the Senate, then, where most members juggle a broader array of constituency interests and where even the most urban states have a farm sector, the formal logrolling role of food stamps in the Farm Bill was more situational than in the House, where the rural members of the Committee on Agriculture were increasingly becoming demographic outliers.[40] As Huelskamp summarizes:

> After 1970, the House committee never tried a farm bill without food stamps. For good reason, it would likely have been a major committee defeat on the House floor. For there are many suburban and urban House districts with few or no farmers; the Senate, on the other hand, contains few states with only limited agricultural interests. Consequently, the Senate panel was apparently less compelled to logroll than its House counterpart.[41]

In 1996 it was new House Speaker Newt Gingrich of Georgia who pushed to move food stamps into a separate bill as part of the House Republican "Contract with America" to reform entitlement programs. However, House Agriculture Committee chair Pat Roberts of Kansas, fearing a fatal loss of support from urban and suburban members of both parties, convinced Gingrich to back a two-year reauthorization of the food stamp program within FAIR, which was enough to overcome defections by farm bloc legislators opposed to controversial "freedom to farm" reforms to commodity programs.[42] The Republican-dominated 104th Congress later made significant cuts in food stamp eligibility as part of a stand-alone welfare reform act signed into law by President Bill Clinton just months before the 1996 presidential election.

Nutrition advocates fought hard against those cuts but found themselves with far less leverage once food stamps were placed in a welfare bill whose beneficiaries tended to have little social and political standing. Their bitter experience in 1996, with both Clinton and congressional conservatives, reinforced their determination to keep food stamps in the Farm Bill and to make food stamps *the* price for supporting it at all.[43] Their message was unambiguous—no food stamps, no Farm Bill—and the Agriculture Committees, eager to retain urban votes, made no effort to decouple food stamps and commodity programs in assembling either the Farm Security and Rural Investment Act of 2002 or the Food, Conservation, and Energy Act of 2008.

Coalitions

The food stamp program became the essential glue binding the tenuous allies in support of the Farm Bill, but it was only one of many deals the farm bloc was willing to make to protect commodity programs, which were facing growing scrutiny because of their costs and their impacts on the food system. However, with each reauthorization, the expanding scope and diversity of the organized interests to be cultivated made it harder for the Agriculture Committees to cobble together their minimum winning coalitions. Refer back to chapter 2's list of titles in the Food, Conservation, and Energy Act of 2008, and you get an idea of the challenges facing Farm Bill architects.

By early 2012, it was possible to think of Farm Bill politics as the positioning of Agriculture Committee leaders like Frank Lucas and Debbie Stabenow to accommodate elements of four distinct if not always unified and sometimes overlapping "advocacy coalitions." These coalitions are defined by Paul Sabatier as loose networks of allies holding similar core policy beliefs, including "legislators, agency officials, interest group leaders, judges, researchers, and intellectuals from multiple levels of government."[44] The types of organized interests composing the four relevant advocacy coalitions are outlined in table 4.2.

The Farm Bloc. This is the "traditional" agricultural lobby, comprising commodity producers, processors, food retailers, farm equipment makers, seed and chemical companies, agricultural finance and insurance firms, rural development advocates, agricultural researchers, local governments, and biofuel promoters. Though generally conservative ideologically, and often in competition with one another over how to divide up Farm Bill spending, the groups making up the farm bloc tend to be pragmatic: make whatever deals necessary to get the Farm Bill through Congress. The farm lobby's closest allies in Congress are members of the House and Senate Committees on Agriculture, the last refuges of a once significant congressional farm bloc. All other things being equal—and in politics, they rarely are—the farm bloc starts each Farm Bill reauthorization with inherent advantages in terms of economic incentives, connections between legislators and voters, access to

Table 4.2 Examples of Groups in Farm Bill Advocacy Coalitions

- The Farm Bloc
  - American Farm Bureau Federation
  - National Council of Farmer Cooperatives
  - National Farmers Union
  - National Corn Growers Association
  - National Crop Insurance Services
  - American Soybean Association
  - National Cattlemen's Beef Association
  - American Sugar Alliance
  - National Milk Producers Federation
- The Hunger Lobby
  - Share Our Strength (Food Banks)
  - Food First
  - National Association of Farmers Market Nutrition Programs
  - Center on Children and Families
  - US Conference on Catholic Bishops
  - Feeding America
  - Oxfam America
- The Food Movement
  - National Sustainable Agriculture Coalition
  - American Farmland Trust
  - Center on Budget and Policy Priorities
  - Farm Aid
  - Slow Food USA
  - Center for Food Safety
  - Food Democracy Now
  - Humane Society of the United States
  - Center for Science in the Public Interest
- Radical Reformers (listed from "Right" to "Left")
  - Club for Growth
  - Taxpayers for Common Sense
  - Cato Institute
  - Heritage Foundation
  - Environmental Working Group
  - Real Food Challenge
  - Small Planet Institute

campaign funds, political leverage, and even cultural identity. After all, who could be against *farmers*?

The Hunger Lobby. This coalition defends food stamps and other government nutrition programs, including aid to food-insecure populations in other countries, and comprises nutritionists, public health and antipoverty advocates, labor unions, religious groups, and other social justice advocates. While its core organizations and activists tend to be ideologically liberal, they are also pragmatic: they support commodity programs in exchange for farm bloc backing on nutrition programs. As such, their leverage depends largely on the votes of urban and suburban liberal Democrats, for whom there is no Farm Bill without food stamps. Nutrition activists are joined, quietly, by major food producers and grocery retailers like Walmart, for whom spending by food stamp recipients is an important source of their monthly revenues and a not-to-be discussed supplement to the low wages of their own workers.[45] On most other issues, food producers and retailers align with the traditional farm lobby.

The Food Movement. According to Michael Pollan, the "food movement" is unified by "little more than the recognition that industrial food production is in need of reform because its social/environmental/public health/animal welfare/gastronomic costs are too high."[46] This loose coalition is composed of critics of the dominant food system, including nutritionists and public health experts; farm organizations such as the National Farmers Union, which tend to represent small family farms; environmental groups; promoters of organic and "specialty" crops; and unions and other labor advocates. Though politically diverse, its component groups tend to support a more active government role in promoting a more sustainable and socially just food system. On many issues—food access, in particular—its members align with the more focused goals of the Hunger Lobby, although they may differ on whether nutrition programs should restrict access to "unhealthy" foods such as sugary soft drinks. In Congress, this coalition also tends to find its greatest support among ideologically liberal Democrats, but it gets the occasional backing of otherwise conservative Republicans for whom small organic farms come closest to mirroring their ideal of Jeffersonian America.

Radical Reformers. This coalition—united predominantly by its harsh critique of the prevailing system—contains some of the most ideologically pure critics of the food system and the Farm Bill. It is largely composed of and driven by "free-market" conservatives who advocate limited government involvement in and spending on agriculture—and any other sector of the economy—as well as reductions in (if not the elimination of) entitlement social welfare programs like food stamps. The most consistent among them would cut spending on commodity programs as well. On this issue, at least, they are joined in spirit (if not in physical terms) by alternative-energy advocates and environmentalists opposed to "market-distorting" and ecologically harmful subsidies for corn ethanol and other biofuels; likewise, their general opposition to subsidies for commodity crops gives them common cause with some elements in the food movement. In some ways, to borrow economist Bruce Yandle's imagery, radical reformers are a classic "bootleggers and Baptists" amalgamation of odd political bedfellows who have nothing in common except a shared desire for change.[47] As we will see, the most zealous elements of this coalition are not inclined to make any deals that continue to prop up what they see as an expensive and dysfunctional farm policy. In Congress, these tend to be the Tea Party Republicans seeking a sharp reduction in, if not the outright elimination of, federal spending on most domestic programs—including commodity supports *and* food stamps.

Overall, then, the story of the past five decades has been one of a shrinking farm bloc making any deal necessary to preserve its federal safety net. As a result, successive versions of the Farm Bill grew longer, more complex, and more expensive over time, as their legislative managers expanded the number and range of "commodities" included in a final package sufficient to gain its acceptance on the House and Senate floors. Programs to subsidize farmland conservation? That's a "commodity," handled by its own subcommittee, no different from food stamps, biofuels, or organic agriculture. So long as every organized group agrees not to oppose any other group's "commodity," putting together the Farm Bill is the legislative equivalent of using a big stapler: the result may not be elegant or internally coherent, but it hangs together.

But what happens when one group refuses to play along and finally obtains enough leverage in Congress to disrupt the "normal" politics of Farm Bill construction? Can the Farm Bill—or any legislation—survive ideological purity? Everyone was about to find out.

5

"Stop the Spending"

Budget Politics and the "Secret" Farm Bill

Nothing will satisfy us but going full steam ahead. You have to understand, the slogan of my campaign was, "Stop the spending." That's a variation on the same theme that every one of us 87 had. . . . We all agree with the "Pledge to America," but there's still the question: Does it go far enough for our liking, for what we were sent to do?
—Rep. Todd Rokita (R-IN)

Weather metaphors are overused in describing political events. However, the 2010 midterm elections cannot help but evoke images of a category 5 hurricane sweeping everything away and remaking the landscape for years to come. In that election, votes of energized Tea Party conservatives propelled Republicans to a net gain of 63 House seats—the largest midterm swing since 1938—taking them to 242 as the 112th Congress convened in January 2011. The party's largest House majority since 1946 included 87 newcomers, several of whom had defeated Republican incumbents in primary elections to make this Republican caucus the most ideologically conservative in history.[1] The new majority installed veteran legislator John Boehner of Ohio as Speaker and promptly laid out its agenda: repeal the Affordable Care Act, slash spending, and rein in the power of the federal government. And, as history would show, if President Obama and congressional Democrats did not go along, the newcomers were prepared to shut down the government until they got their way.[2]

The election was especially painful for House Democrats from the Farm Belt, who had picked up critical seats in states such as Indiana, Iowa, and Kansas over the previous two election cycles, only to be nearly wiped out in

2010. In 2008, with the Democrats in control of both chambers, Congress had pulled together a Farm Bill that distributed benefits to agricultural interests while keeping urban liberals in the fold by expanding a newly relabeled SNAP. Indeed, stitching together the Food, Conservation, and Energy Act of 2008 had been classic Farm Bill politics: offer a bit of something for everyone, from corn ethanol to organic crops; expand nutrition programs; and staple the final package together. The only ones unhappy with the outcome were fiscal conservatives, who in 2002 had criticized President George W. Bush for not opposing a similarly generous reauthorization. In 2008 Bush vetoed the omnibus package as too expensive, only to be summarily overridden by large bipartisan majorities.

For all their largesse to farmers, Democrats got their heads handed to them in November 2010, with rural voters ousting an already dwindling cadre of conservative Democrats in favor of even *more* conservative Republicans. "What happened?" House Speaker Nancy Pelosi asked House Agriculture Committee chair Collin Peterson. "We gave the farmers everything they wanted." With the highest commodity prices in forty years, Peterson explained, farmers and other rural voters focused less on economic concerns and more on cultural issues such as abortion and same-sex marriage; added to this was their general antipathy to federal regulation, the Affordable Care Act, spending on "undeserving" populations like immigrants, and, to no small extent, Barack Obama himself. So they voted Republican.[3]

In doing so, rural voters helped guarantee that the Farm Bill came up for renewal in 2011 under dramatically altered political circumstances. For one thing, Obama was still president, and the Democrats retained a 53–47 advantage in the Senate. Splits in partisan control, combined with the widest ideological distance between the parties since the 1890s, threatened to make the compromises needed to pull together a new Farm Bill by September 2012 harder to come by.[4] Second, even if the Farm Belt was enjoying its greatest prosperity in memory, the rest of the nation was still suffering the effects of the economic crisis that had propelled Obama to the White House in the first place. The fiscal costs of bailing out the nation's banking system, saving its beleaguered domestic auto industry, and stimulating the economy through government spending had caused a sharp spike in the federal budget deficit, the issue Tea Party activists highlighted in helping Republicans take over the House.

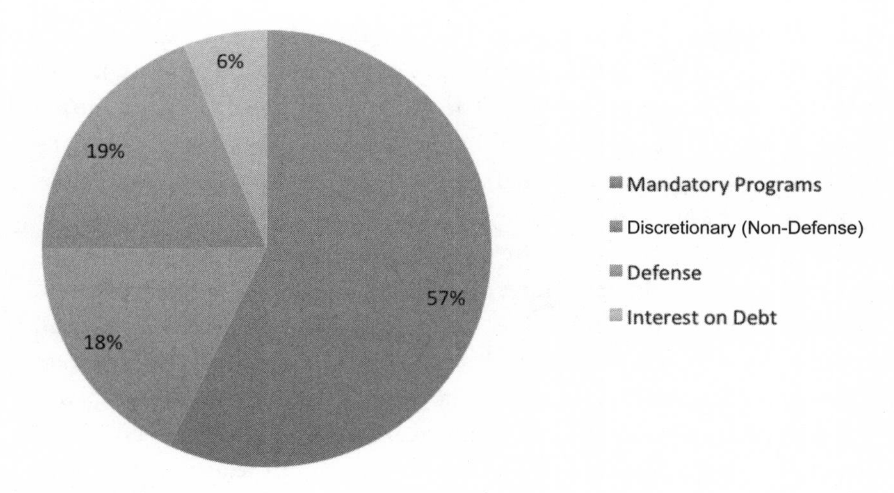

Figure 5.1 The Federal Budget, January 2011
Source: Calculated from *The Budget and Economic Outlook, Fiscal Years 2011 to 2021* (Congressional Budget Office, January 2011), 6, table 1-2.

At the top of conservatives' priorities—right after repealing the Affordable Care Act—was to make good on their "Pledge to America" to cut "discretionary" federal spending back to pre-2008 levels, starting with *immediate* cuts of $50 billion to $60 billion for the remainder of fiscal year 2011.[5] As figure 5.1 shows, "discretionary" was understood to mean (among Republicans, at least) everything *but* national defense, Social Security, Medicare, and other "entitlement" programs (which, by the way, included SNAP)—in other words, roughly 18 percent of a $3.7 trillion federal budget that encompassed everything from education and environmental regulation to housing, food inspection, and, of course, agriculture. To get there would require an immediate 30 percent cut in these programs—and more in years to come.[6] As one veteran Republican later observed, "the entire emphasis of so many members was 100 percent deficit reduction come hell or high water. There were no real substantive conversations about anything but cut, cut, cut, cut, cut."[7] Their drive to cut would shape the overall agenda for everyone.

The new House leadership started by banning "earmarks," a cherished tradition by which legislators distribute benefits to constituents by inserting targeted spending provisions into authorizing and spending bills.[8] Such directed spending can vary widely, from new highway bridges and research buildings at local universities to wind-power demonstration projects, homeland security grants, and agricultural pest inspection stations. Indeed, during negotiations over the previous two Farm Bills, House Agriculture Committee chairs of both parties had used earmarks to build support among fellow committee members, distributing targeted funds for biomass energy research, rural economic development, cattle waste management, food safety, and barley gene mapping, among many others.[9] Truth be told, any ban on earmarks would have a trivial effect on overall federal spending, but it had considerable symbolic value insofar as it signaled an end—in the House, at least—to traditional "pork barrel" politics and its image of wasteful spending. However, to many veteran Republicans and other defenders of earmarks, they were the lubricants that helped "grease" the legislative process.[10] Without earmarks, making deals to build support would be harder. But the leadership held firm: no more special goodies.

In fact, Boehner went further and declared his intent to end omnibus bills altogether. Rather than allowing the sprawling "catch-all" bills that had come to typify legislating—the Farm Bill included—the Speaker stated his preference for smaller, more focused measures that members would vote for based on their purported merits, not because the final package happened to contain a desired benefit for some organized interest or for the folks back home. That Boehner had voted against the Farm Bill in both 2002 and 2008 lent credibility to his vow.[11]

Finally, House leaders pledged there would be no new spending without offsetting cuts elsewhere, although they pointedly exempted any additional costs incurred in a successful repeal of the ACA. More important, they enabled the Budget Committee, now chaired by Paul Ryan of Wisconsin, to impose limits on what the authorizing and appropriations committees *could* spend. Want to increase spending for commodity supports, agricultural research, or nutrition programs? What are you willing to cut to find the offsetting funds within your allotted spending?

Together, these rule changes shifted power in the House away from authorizing committees like Agriculture and even from the Appropriations Committee, which normally decides annual spending within the boundaries set by the authorizing committees. Effective power over spending priorities was now in the hands of Budget Committee chair Ryan, who also gained the authority to unilaterally impose spending limits on other committees until Congress passed a fiscal year 2012 budget by the September 30 deadline.[12] In effect, Ryan, who had also voted against the Farm Bill in 2008, now had Boehner's blessing to tell other committee chairs how much money they had to play with.

The 112th Congress would be dominated by a bare-knuckles, high-stakes fight between congressional Republicans and the Obama administration over the federal budget—and, in some respects, the very role of the federal government—with congressional Democrats working to defend the president's priorities. In 2011 that fight would lead to successive threats by House Republicans to shut down the federal government, the creation and eventual failure of a House-Senate "supercommittee" to come to a mutually acceptable long-term budget plan, and, in the House at least, mandates on what the authorizing committees could spend in future years.

As a result, and reminiscent of when Congress had struggled to pass what became the Federal Agricultural Improvement and Reform Act of 1996, leaders of the House Agriculture Committee faced the prospect of losing control over their own agenda as they sought to reauthorize the Farm Bill by the September 30, 2012, deadline. Under orders to slash spending, Lucas and Peterson would be presented with a cleaver—their new figurative tool for Farm Bill construction. Both knew it would have been easier and more pleasant to staple together a package of distributed benefits for those members whose votes were needed to gain passage. By contrast, the specter of cuts would pit winners against losers, making it that much harder to get everyone to yes. But that was the task they faced.

The Waiting Game

Everyone with an interest in agriculture and nutrition wanted to know one thing when the 112th Congress convened: when would the Agriculture Committees start work on the Farm Bill? It soon became clear that neither Frank Lucas nor Debbie Stabenow was in a hurry. For one thing, the reshuffling of partisan control in the House and prospects for a seismic clash over the federal budget had so unsettled matters that neither chair thought it wise to move until the broader spending contours were clearer. As Senator Stabenow put it, "In terms of an exact timetable, I think it's in the interest of agriculture, given the backdrop that we're in right now around budgets and deficits, to move in a thoughtful, methodical way and get this in place as soon as we can."[13]

More problematic for Lucas and Peterson on the House side, the 2010 elections had brought in a raft of new members whose links to agriculture differed widely from their own (see table 3.2). Reflecting their party's losses in the Farm Belt, committee Democrats welcomed only one newly elected member: Terri Sewell of Alabama's Seventh District, a majority African American district centered around the city of Birmingham. Sewell cared about nutrition programs, not cotton. Similarly, most returning committee Democrats hailed from metropolitan areas such as Cleveland and Boston or, like Chellie Pingree of Maine, from areas in the Northeast and on the West Coast where farming issues were very different from those in the midwestern and southern bastions of commodity agriculture. For his part, Lucas had to assimilate into his committee fifteen new Republicans, including Tim Huelskamp of Kansas, almost all of whom identified as Tea Party conservatives whose stated priority was cutting federal spending above all else. Lucas soon made it clear that he would use 2011 for a series of "audit hearings" to educate these rookies on agricultural policy and to identify areas for budget savings. He wasn't likely to start work on the Farm Bill until early 2012, regardless of what farm groups wanted. The waiting game had begun.

However, the absence of formal committee deliberations should not be confused with an absence of activity. While the various coalitions of organized interests positioned themselves for whenever the committees might start formal consideration, higher-level dynamics within Congress were shaping the terms under which those committees would eventually approach reauthorization. Chief among them was the fight over the federal budget.

Sacrificing Direct Payments

The battle of the budget began in late January with President Obama's 2011 State of the Union address, in which he defended his record and, in a bit of positioning for the 2012 election, proposed a freeze on nondefense discretionary spending that would save about $400 billion over ten years. Republicans derided Obama's proposal as inadequate, and Boehner reiterated their pledge to cut spending to 2008 levels, starting with $100 billion in *immediate* cuts.[14]

These competing proposals might have been more theoretical—trial balloons for the fiscal year 2012 budget, as it were—had the 111th Congress actually passed a fiscal year 2011 budget. But it had been locked in a preelection struggle over spending, with Senate Republicans using their chamber's rules to block the budget bill from a full vote, so lawmakers had approved funding only through March 4, 2011.[15] As a result, the 112th Congress would need to authorize spending for the remainder of the fiscal year (through September 30, 2011) or face shutting down "nonessential" federal programs. That window of opportunity gave Republicans critical leverage as the new Congress convened, and Boehner pledged that Republicans would not support any short-term spending bill without significant additional and immediate cuts.[16]

Any new cuts would obviously affect the agriculture and nutrition programs under the Farm Bill, whose costs by now had risen to approximately $100 billion a year. More critically, spending on SNAP had skyrocketed to nearly $75 billion a year with the Great Recession, and in the eyes of House Republicans, the program was ripe for reforms and cutbacks. However, they knew that cuts in SNAP spending had to be accompanied by cuts in farm subsidy programs if they were to have any chance of gaining support in the Senate, much less with President Obama. Attention quickly turned to the $10 billion to $30 billion in annual support for commodity production.[17] To critics along the ideological spectrum, from the Environmental Working Group on the Left to the Cato Institute on the Right, this array of direct payments, subsidized crop insurance, and marketing loans amounted to little more than welfare for producers who didn't need the help. Their greatest scorn was aimed at the $5 billion in direct payments that went to recipients regardless of income level—and regardless of whether they planted a crop. That program, devised by House Agriculture Committee chair Pat Roberts as part of the 1996 bill, was always defended as a safety net against inevitable bad times. However, program costs had grown over the years, and revelations of payments made to affluent absentee landlords—some in Manhattan, New York, *not* Manhattan, Kansas—made it a symbol of misplaced priorities. Despite such criticism, Roberts and other farm bloc legislators, backed by the American Farm Bureau Federation and most commodity groups, had fended off efforts to impose income limits on recipients during both the 2002 and 2008 reauthorizations.

In 2011 the politics of the budget gave critics the advantage. The first sign came in February from Obama, whose fiscal year 2012 budget proposed income caps on direct payments to farmers and trimmed subsidies on crop insurance, all to save roughly $5 billion a year.[18] Democrats, stung by their 2010 losses, no longer seemed so concerned about alienating Farm Belt voters. Big-city newspapers inveighed against the program, with the *New York Times* opining, "here is one big-ticket saving that all members of Congress should get behind: cutting the billions of dollars in farm subsidies that distort food prices, encourage over-farming and inflate the price of land."[19] For his part, House Budget Committee chair Ryan declared, "We shouldn't be giving corporate farms, these large agribusiness companies, subsidies." Ryan's eventual spending plan proposed to cut $48 billion from farm programs over ten years, to go along with significant cuts in spending for nutrition programs.[20] Speaker Boehner, never a fan of commodity programs to begin with, labeled direct payments a "slush fund."[21]

This convergence of criticism sent a clear message to farm groups and their congressional patrons: cuts in commodity programs were inevitable, so set priorities. Not surprisingly, members of the Agriculture Committees were wary of changing, much less eliminating, a direct payment system that worked reasonably well for a range of crops. Chairman Lucas, whose childhood memories included a hailstorm devastating his father's wheat crop, called Obama "out of touch with agriculture" when the president proposed that direct payments go only to farmers with less than $500,000 in annual sales.[22] Members from southern rice, peanut, and cotton areas were particularly opposed to losing direct payments, arguing that crop insurance did not work for their crops and smaller-scale producers. In March, Lucas and Peterson proposed to Ryan that any additional cuts in Farm Bill spending for the remainder of fiscal year 2012 come entirely out of the nutrition title, arguing that direct payments were an essential safety net to protect against the "inevitable" fall in commodity prices.[23] Behind their request was a broader message: let the Agriculture Committee set priorities.

But it was clear even to Farm Belt legislators that something other than nutrition programs had to be cut. As noted, cuts that focused solely on SNAP would never get past the Senate, much less Obama. Moreover, high commodity prices and generally strong farm incomes made direct payments easy targets, particularly when compared with the relatively small amounts going to individual SNAP recipients. "The scrutiny of farm programs is stronger than ever," admitted Chuck Conner, president of the National Council of Farmer Cooperatives.[24] The first to break from the past was the Iowa Farm Bureau, a major voice in corn and soybeans; in early 2011 it voted to end direct payments in exchange for expanded crop insurance coverage, which worked well for Iowa corn and soybeans. Other farm and commodity groups, and finally the national AFBF itself, fell in line as the year progressed. In their view, it was better to give up or cap a benefit that relatively few needed—and that many outside agriculture derided as welfare for the rich—than to face steeper cuts in programs of greater overall importance, such as subsidies for crop insurance premiums.[25]

By midyear, as the budget battle intensified, even members of the Agriculture Committees saw that direct payments were no longer worth defending. Lucas and Peterson may have still regarded them as a safety net, but to the many newcomers on the House committee, cuts in farm subsidies fit within their overall attack on federal spending.[26] Tim Huelskamp, whose First Kansas had received $250 million in direct payments in 2010—the second highest level in the country—told his constituents that defending such subsidies was hard when so many other programs were facing cuts. "Farmers are going to have to make the argument, to Head Start folks and others, that their subsidies are worth borrowing 42 cents for every dollar spent," he said.[27] In the Senate, while Pat Roberts continued to defend the program he had devised two decades earlier, Debbie Stabenow, in her first months as chair, was more circumspect: "We know there are changes that have to be made. We have to be credible, frankly, in going forward. When we talk about the safety net, in order to be credible, we're going to have to work together to make changes, but it should be the Agriculture Committee doing that, not people who don't understand agriculture."[28] Whether she was issuing a challenge or making a plea to the budget cutters was unclear.

The Semisecret Farm Bill

In early April 2011 Congress sidestepped a shutdown of "nonessential" federal services by agreeing to a budget deal that cut an additional $37.8 billion in nondefense discretionary spending through the end of September, amounting to $78 billion in one-year spending cuts overall. This agreement was $7.8 billion more than the Democrats wanted, but far less than the $61 billion in cuts sought by House Republicans—fifty-nine of whom pointedly refused to support a deal worked out by their own leaders.[29] Any relief was short-lived, as yet another deadline loomed: Congress had to vote to formally raise the nation's $14.3 trillion borrowing limit. If it did not do so, the Treasury Department would be barred from standard-practice borrowing to pay bills (it normally keeps enough cash on hand for only a few months of spending), technically causing the nation to default on its debts. Over the previous century, votes to raise the debt limit had been largely routine, as few in either party wanted to risk the nation's creditworthiness. The telling exception came in 1995 when the Republicans, newly in control of Congress, came close to breaching the limit but instead instigated two partial federal government shutdowns in an ultimately unsuccessful and politically costly fight with President Clinton. Even with that experience, most conservative Republicans in 2011 saw raising the debt limit as enabling what they considered unsustainable borrowing to cover gaps between revenues and expenses.[30] Given their pledges *never* to raise taxes—a vow enforced by well-funded conservative activist groups and radio talk-show hosts on constant alert for ideological apostasy—the new House majority was determined to use the debt limit to force the spending cuts its members had promised.

Seeking to avoid an impasse on the debt limit and hoping to settle the budget issue before the 2012 election, President Obama proposed a "grand bargain" to balance the annual federal budget within ten years. It involved cutting $1 trillion in domestic and defense spending; cutting $650 billion in future spending on Medicare, Medicaid, and Social Security; and producing $1.2 trillion in additional revenues by allowing Bush-era tax cuts on higher-income earners to lapse at the end of 2012.[31] Although his proposal got some bipartisan support in the Senate, where a plurality of Republicans was lukewarm to the conservatives' "hostage-taking" strategy on the budget, for House Republicans, it was dead on arrival because it included "new" tax revenues. Boehner and Ryan came up with a counterproposal to raise the debt limit by $900 billion through April 2012, in return for $917 billion in new budget cuts, with no new revenues. However, they were unable to get the support of the most conservative members of the Republican caucus, led by majority leader Eric Cantor of Virginia, because their plan did not cut deeply enough.[32] Boehner, whose speakership hinged on the support of the caucus, and mindful that Cantor wanted his job, offered a plan in late July to cut $5.8 trillion over ten years, double the amount under discussion with the White House. Few outside observers saw Boehner's proposal as serious, given a promised presidential veto, but it gained the approval of House conservatives eager for a showdown on the budget. The House passed Boehner's proposal by a party-line vote. It subsequently went down to defeat in the Democratic-controlled Senate, also by a largely symbolic party-line vote.

In August, with a federal government shutdown hours away and with the nation's creditworthiness having been downgraded for the first time in its history, House and Senate leaders agreed to a last-minute deal to extend the debt limit through January 2013 and to create a House-Senate "supercommittee" to recommend ways to reduce the federal budget deficit by at least $1.2 trillion over ten years. If the Joint Select Committee on Deficit Reduction, as it was formally called, failed to agree to a plan that could gain approval in both chambers by November 23, the agreement called for an automatic "sequester" of $900 billion in across-the-board budget cuts over ten years, beginning in January 2013. Such a sequester would affect all nonentitlement spending, defense included, in an effort to spur compromise among all involved.[33] Despite opposition by many conservative Republicans and liberal Democrats, for opposite reasons, both chambers approved the plan, thereby avoiding technical default.[34]

Creation of the supercommittee, chaired by Vice President Joseph Biden, had immediate ramifications across the federal establishment because any outcome, even the threatened automatic cuts, would shape federal spending priorities for years to come. Most critically, supercommittee negotiators were given the freedom to spell out cuts *and* changes in specific programs, effectively overriding congressional authorizing and appropriations committees. At the same time, the process offered opportunities for members of Congress who were not on the supercommittee to quietly convince their colleagues who *were* on it where to cut—and where not to cut—knowing that any final deal was subject to a straight up-or-down vote, similar to when Congress approved conference committee reports. Supercommittee members became *very* popular.

Leaders of the House and Senate Agriculture Committees were notable among those negotiating behind the scenes. They saw this special panel as a mechanism for setting farm and nutrition program spending priorities without having to go through a protracted, highly public fight over reauthorization in a context defined by cuts. In October, Stabenow, Lucas, Roberts, and Peterson recommended a package to the supercommittee that contained $23 billion in cuts over ten years, with $14 billion coming from commodity programs, $6 billion from conservation programs that were being underutilized as farmers kept land in production to take advantage of high commodity prices, and the remaining $3 billion from nutrition programs, largely in the form of tightened SNAP eligibility requirements.[35] But their recommendations also included fundamental policy changes in the commodity title, chief among them a proposal to replace direct payments with a plan that gave farmers a choice between a risk coverage program for losses not reimbursed by crop insurance (so-called shallow loss coverage) and, to satisfy southern rice and peanut growers, a version of a pre-1996 countercyclical program based on crop trigger prices first included in the 2002 bill.[36] Finally, this proposal would reauthorize the Farm Bill's commodity title—its heart—for another five years.

Other key farm bloc legislators, led by Senate Budget Committee chair Kent Conrad (D-ND), were receptive to the strategy, seeing budget negotiations as an opportunity to expedite commodity program reauthorization under special rules that required only fifty votes on the Senate floor (Biden would presumably break a tie). By contrast, a stand-alone Farm Bill would be considered under the normal rules, which would essentially require sixty votes to end debate in the likely event that one or more senators objected to some provision. To their way of thinking, the supercommittee offered a way to extend farm and nutrition programs for at least five years, with regrettable but manageable cuts, while avoiding possibly worse outcomes if the Farm Bill went through the normal process.[37]

If that strategy appealed to farm bloc leaders seeking to maintain control in unsettled political and fiscal conditions, rumors of a "secret" Farm Bill caused consternation among the many advocacy groups that had been waiting to have their say about reauthorization. Commodity groups such as the National Corn Growers Association and the National Cotton Council, though unhappy at the prospect of cuts and changes in subsidy programs, at least trusted Agriculture Committee leaders to protect the core elements of established policy. But other advocacy groups expressed outrage that a closed-door budget process would essentially define food policy priorities without public hearings or opportunities to propose amendments. Critics of the existing farm support system, ranging from economists in "free-market" thinks tanks like the American Enterprise Institute to environmentalists in the Environmental Working Group, attacked the prospect of supercommittee members simply replacing one set of unjustifiable commodity supports with another.[38] Food system activists in groups such as the National Sustainable Agriculture Coalition, who had been rallying behind Chellie Pingree's Local Farms, Food, and Jobs Act, were especially upset at the prospect of a "backroom" deal. "Right now 4 members of the House and Senate ag committees are meeting to rush the 2012 Food and Farm Bill to the Super Committee and steal any chances for reforms for local, organic and healthy food until the next Farm Bill comes up in 2017," Food Democracy Now complained to its supporters, who responded by making 27,000 phone calls to members of Congress in protest. At the same time, a group of two dozen House members, led by Democrat Ron Kind of Wisconsin, submitted a letter to the supercommittee urging it not to consider changes in farm programs outside the normal process.[39] But the central players soldiered on, hoping to lock in what they saw as reasonable adjustments, given the political and fiscal uncertainty.

In the end, however, the supercommittee could not reach an agreement, largely due to House Republicans' intransigence over any efforts to raise revenues. The collapse of the supercommittee effort meant that the first "sequester" of automatic cuts would take place in January 2013. But the House and Senate, after yet another brush with a government shutdown, agreed in mid-December to a $915 billion spending package to the end of fiscal year 2012. Passage in the House (296–121) came over the objections of more than 80 Republicans who wanted steeper cuts.[40]

Imposed Boundaries of Farm Bill Negotiation

When the supercommittee failed, so did the efforts by Agriculture Committee leaders to use the special budget process to their advantage. Yet, in key ways, the broad contours of the new Farm Bill had already been set by the time the House Agriculture Committee held its first field hearings in March 2012. For one thing, the reality of spending cuts made it easier to get farm groups to acquiesce to replacing the much criticized system of direct payments with an expanded form of subsidized crop insurance. Whereas Senator Stabenow called it a "monumental shift in federal farm policy" that "saves billions of taxpayer dollars by ending payments to farmers who don't need them,"[41] agricultural economists doubted that a "backdoor" subsidy of essentially free crop insurance would prove less expensive over the long term, especially if drought or extreme weather events increased insurance payouts.[42] Even so, all agreed that moving to a system of insurance, even if subsidized, would make crop supports less politically exposed in the coming reauthorization battle. After all, who could be against insurance?

Equally important, it was clear to all that Farm Bill reauthorization in 2012 would start from the premise of less, a profound change from 2002 and 2008. *How much* less, and *for whom*, was still to be determined, as House chair Lucas acknowledged in opening the first field hearing in upstate New York:

> We also will be spending less money on the next farm bill, whether it's the $23 billion reduction in spending compared to the previous farm bill that was agreed to by the principals of the Agriculture and Senate Committee or the President's $32 million proposed reduction, or the $40+ billion reduction suggested last year by the House Budget Committee, we'll have less money to spend. So that makes our challenges tougher trying to be responsible and keep the good things.[43]

If Lucas was intentionally speaking with understatement, even he had no inkling how hard his task was about to get.

Figure 6.1 Timeline of Farm Bill, 112th Congress

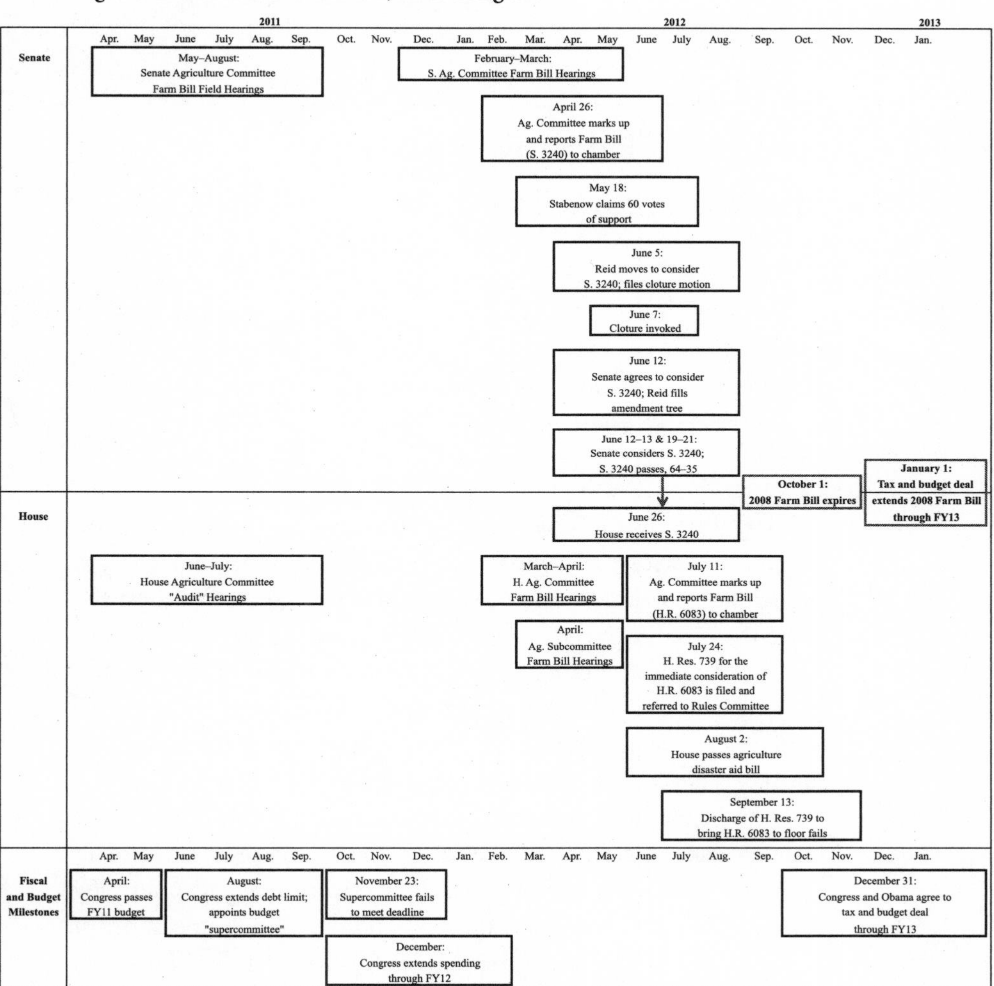

6

Building a Pathway to Sixty

The Senate Moves First

If somebody has a problem with the bill, come to us.
—Sen. Debbie Stabenow

In early 2012, in the wake of the collapse of the supercommittee process and with presidential and congressional elections fast approaching, it was not clear that Congress was going to act on the Farm Bill—or much else, for that matter. Longtime agricultural policy observers were hard-pressed to figure out who in Washington had any incentive to push reauthorization, as opposed to simply extending the existing bill beyond the fall elections, given that either effort was likely to get mired in ongoing deficit politics.

In the Senate, Debbie Stabenow was facing what looked to be a tough general election challenge by former representative Peter Hoekstra, and some experts thought she would benefit politically if she could maneuver a bill through committee but stop short of a potentially rancorous and unsuccessful floor battle. Many also wondered whether Pat Roberts, Stabenow's ranking minority colleague, might prefer to wait until 2013, in the hope that Republicans regained the Senate and he could take over as committee chair and thereby aid his own reelection in 2014. In the House, few thought the Democrats would be back in charge anytime soon, so the question facing Frank Lucas was how to gain the support of the many first-term conservative Republicans whose zeal for fiscal austerity outmatched his own.

Finally, there was the issue of the president's reelection: was it in Obama's interest to push action on the Farm Bill in 2012?[1] Although he had done reasonably well with rural voters in 2008, compared with previous Democratic candidates John Kerry and Al Gore, his election had depended overwhelmingly on urban and suburban votes. For their part, rural voters had played an outsized role in propelling the Tea Party into prominence within the Republican Party, largely at the expense of rural Democrats. With the farm economy doing better than the nation's economy overall, few expected the president to expend political capital on the Farm Bill without clear expressions of support from the nutrition and food policy reform activists in his base.

Putting off an effort to reauthorize until 2013 had some advantages. A major question for everyone was whether the Farm Bill would fare better in 2012, given the prospects of at least $23 billion in cuts—the amount pledged to the supercommittee—or in 2013, even if obtaining a simple extension beyond the September 30, 2012, deadline opened commodity programs to the possibility of $10 billion to $15 billion in automatic cuts if the January 2013 sequester went into effect. Collin Peterson mused that he would "much rather pass a baseline bill and take our chances under sequestration."[2] His lack of alacrity was shared by nutrition program advocates; sequestration would exempt entitlements such as food stamps, while any effort to push for a bill in 2012 opened up the possibility of at least $4 billion in cuts per the proposal to the supercommittee, and much more if Budget Committee chair Paul Ryan and other House Republicans had their way.[3] Nutrition advocates, along with food activists, most of whom supported Obama's reelection, also hoped that the fall elections might strengthen Democrats' position in Congress.

By contrast, farm groups wanted reauthorization sooner rather than later, with commodity producers in particular seeking certainty on the form and scope of crop support programs. Farm groups like the American Farm Bureau Federation also were wary about their ability to stave off the effects of continued pressure from the Republican base to cut the budget. For them, as Farm Bureau lobbyist Mary Kay Thatcher commented, there was "no upside" to waiting until 2013.[4]

Regardless of the opinions of various outside interests and prognosticators, all the central players involved publicly expressed their intent to act before the 2008 law expired on September 30. Secretary of Agriculture Thomas

Vilsack made it clear that the president was ready to work with Congress.[5] House and Senate Agriculture Committee leaders separately stressed that they wanted to get a bill done in 2012, with Lucas expressing concerns about the electoral impact of inaction on rural House Republicans.[6] Moreover, even though the supercommittee process had been a failure in the strictest sense, it had given Lucas and Stabenow a head start in marking up a consensus bill that, with some tweaks, might move quickly through their respective committees. On April 22 the two chairs announced that they would produce a "joint mark" to be presented to the Senate Agriculture Committee on April 25. The waiting game seemed to be over.

First-Mover Advantage

Congress is a true bicameral legislature—its two chambers are coequal—but does it matter which goes first in kicking off the legislative process? Article I, section 7, of the Constitution mandates that bills raising revenues must originate in the House of Representatives. Otherwise, it is silent on the matter. This does not mean that which chamber goes first isn't important. If nothing else, the "first mover" sets the agenda, in tone and in substance. As students of policy change know well, "*when* things happen in a sequence affects *how* they happen."[7]

But *why* did the Senate Committee on Agriculture, Nutrition, and Forestry move first in considering S. 3420, the Agriculture Reform, Food, and Jobs Act of 2012? History offers no real guide. In fact, the House Committee on Agriculture had been first out of the gate on the three previous reauthorizations of the Farm Bill. The difference this time around may have been the starkly disparate levels of experience on the respective panels, with Senator Stabenow having the good fortune to chair a committee of seasoned legislators who had worked on other farm bills. More to the point, five of the committee's twenty-one members had actually chaired the panel at one time or another: Democrat Patrick Leahy of Vermont had led the committee during the 1990 reauthorization and could speak to the concerns of Northeast agriculture, dairy in particular, and could be counted on to defend SNAP; Democrat Tom Harkin of Iowa, a powerful voice for Midwest agriculture, had led it during the 2002 and 2008 reauthorizations. Republicans Saxby Chambliss of Georgia, a former agricultural lawyer with expertise in peanuts and cotton; Thad Cochran of Mississippi, who could represent southern agriculture while cutting the deals needed to get a bill passed; and Richard Lugar of Indiana, a leader on biofuels research and school nutrition programs, had led the committee through the 1996 reauthorization.[8] Additionally, ranking minority member Pat Roberts had chaired the House committee in 1996, and another Republican, Michael Johanns of Nebraska, had served as secretary of agriculture for two years under George W. Bush. Given such experience and expertise among her colleagues in both parties, it was with some legitimacy that Stabenow could declare: "This is not a partisan committee. This is a committee where we focus together on farm policy. We may have differences about which crops we advocate for or what we believe is the most important focus, but it is very much done on a bipartisan basis."[9]

In part because of their experience, Senate committee members also enjoyed considerable influence with their colleagues in the larger chamber. Leahy and Harkin in particular could be counted on to build support among Democrats, while Roberts and Lugar could do so among Republicans. Other key members included Budget Committee chair Kent Conrad (D-ND), who had a reputation for consensus building even while defending the needs of the Plains states, and Senate minority leader Mitch McConnell (R-KY). McConnell could enforce any procedural deals worked out with majority leader Harry Reid (D-NV), despite tensions over the former's effective—and, in Reid's view, infuriating—use of Senate rules to delay and obstruct President Obama's legislative agenda from the moment he entered the White House.[10]

By comparison, the House Committee on Agriculture chaired by Frank Lucas stood out for its *lack* of experience, especially among the many newly elected Republican members. As a result, Lucas seemed content to let Stabenow and the Senate take the lead in moving on the plan devised during the supercommittee process; meanwhile, he continued to hold hearings to educate his members on the essentials of agricultural and nutrition policy. Perhaps more critical, Lucas likely knew that his committee lacked the influence in the larger chamber enjoyed by its Senate counterpart. Even the most urbanized state still had an agricultural sector worth some attention by its two senators; in contrast, comparatively few single-member House districts were dependent on farming, making the rural members of the House Agriculture Committee more of a demographic outlier with each passing election. "Given this over-representation of rural interests on the House panel," Tim

Huelskamp had observed back in 1996, "committee decisions on farm bills in the House are more likely to be challenged at the chamber level than in the Senate."[11]

The potential for such challenges were even greater in 2012: On one side of the party divide, the near obliteration of rural Democrats in 2010 had severely eroded Collin Peterson's leverage in his largely liberal, largely urban and suburban party caucus. On the other side of the aisle, Frank Lucas could not move until he knew whether Budget Committee chair Ryan and the Republican leadership would go along with the $23 billion in cuts that had been agreed on by the four Agriculture Committee leaders during the supercommittee process. Nor did he know whether Speaker Boehner would allow a bill to go to the floor without clear signals that it would be supported by conservatives intent on reining in federal spending, farm and nutrition programs included. The literal collapse of an ideological center in the House would severely constrict the Agriculture Committee's room to maneuver.

Gathering Votes

Stabenow was well positioned to move first. Sensing that she had a coalition of support following negotiations over the "secret" bill during the supercommittee process, Stabenow announced that the committee would mark up a bill in late April, using the joint bill worked out with her House counterparts as a starting point. But that version, reflecting the efforts of four Agriculture Committee leaders responding to a budget mandate, left three major challenges: commodity interests, food stamps, and milk.

Balancing Commodity Interests. The primary challenge was to maintain the critical balance among regional crop interests. This task, always at the top of Farm Bill negotiators' list of concerns, was harder this time around because it was clear that Congress would not keep the much-maligned direct payment system.[12] In other years, Stabenow might have retained elements of the old system to ensure the votes of southern senators concerned about rice, peanuts, and cotton, for which countercyclical payments had been included in the supercommittee deal worked out the previous fall. By April 2012, however, such payments were politically unpalatable to a majority of senators in both parties. More important, Stabenow could show promised budget savings with the new "shallow loss" insurance program, whose lower up-front costs "scored" better in Congressional Budget Office (CBO) assessments of the bill's five- and ten-year budget impacts. Left unstated were any costs that might be incurred later, should drought or other "unexpected" weather disasters cause insurance premiums and payouts to skyrocket.[13] But assumptions about up-front costs mattered most, and direct payments fared badly by comparison.

Moreover, putting some form of direct payment back into the mix would mean cutting somewhere else—nutrition programs, most likely—to stay within the $23 billion target. Deeper SNAP cuts were nonstarters with most of Stabenow's Democratic colleagues, not to mention with the White House, so one idea floated during committee hearings in March was a streamlined direct payment program with far higher target prices that would score better in CBO assessments. Roberts seemed open to the idea, which also had some support among rice and peanut growers; however, it was opposed by most corn, soybean, wheat, barley, and oat growers, who complained that it would complicate their planting decisions.[14] At the same time, Stabenow had to handle objections by committee members from the Northern Plains—Democrats Kent Conrad of North Dakota and Max Baucus of Montana, and Republicans John Hoeven of North Dakota and John Thune of South Dakota—who complained that neither shallow loss coverage nor higher target prices worked for their region's producers of livestock and trees in particular.[15] They wanted targeted catastrophic coverage for those crops, but where to find the money to pay for it was anyone's guess. Finally, for every member, the strategic question was whether to try to resolve these divergent commodity concerns in committee, push them off to the Senate floor, or wait for some eventual House-Senate conference committee.

Food Stamps. The joint mark called for a $4 billion reduction in SNAP spending over ten years, largely by closing a "loophole" whereby recipients in a number of northern and Pacific coast states were eligible for higher benefits if they received as little as $1 a year from the federal-state Low Income Home Energy Assistance Program (LIHEAP). To critics, this low LIHEAP threshold demonstrated that SNAP was out of control, and some Republicans wanted to prohibit states from considering LIHEAP in calculating SNAP benefits at all. SNAP advocates argued that eliminating or tightening the LIHEAP link punished deserving recipients, especially the elderly, who might be forced to cut back on food purchases to heat their homes.[16] For committee Democrats, raising the LIHEAP threshold to $10 a year was a regrettable but modest trim of what was now a $75 billion annual expense. Their argument was amplified by actions in the House, where Ryan had just released a proposed ten-year budget calling for *$133.5 billion* in cuts in nutrition programs as part of a $180 billion reduction in overall Farm Bill spending, $31 billion of which came from commodity programs.[17] Although Ryan's plan prompted the House Agriculture Committee to pledge to reduce SNAP spending by $35 billion over ten years, in the hope of staving off even deeper cuts, few thought that either proposal had any chance of success in the Senate, much less the White House.[18] Even so, threats of drastic cuts from the House bolstered Stabenow's argument that nutrition program advocates could live with a $4 billion reduction over ten years.

Got Milk? The joint mark sought to at least partially address what had become an increasingly bitter battle within the dairy industry. Since the 1938 Agricultural Adjustment Act, milk production had been structured by a dizzying array of regional and subregional "marketing orders" that set minimum bulk milk prices for different parts of the country, each based on local production conditions and costs. Smaller producers, mostly in the Northeast and upper Midwest, had suffered for years due to sagging bulk milk prices and rising feed costs, and they sought to create a new milk "stabilization" program to lessen the gap between income and expenses. Their needs were embodied in a bipartisan Dairy Security Act introduced in the House in 2011 by Collin Peterson and Mike Simpson (R-ID), and provisions from this bill were now inserted into the joint mark's commodity title. However, the plan was opposed by larger dairy operations, located mostly in western and southwestern states, for which the existing system worked well, and by milk processors, for whom "management" meant higher bulk milk prices, lower profit margins, and, they argued, higher costs to consumers.[19]

Stabenow knew the committee would be unable to resolve all these conflicts and still meet the goal of $23 billion in cuts. She also knew that her ultimate objective was not the eleven votes needed to report the bill out of committee but sixty votes—the number needed to cut off floor debate should any senator threaten to stall consideration with a filibuster. So, in an example of classic Farm Bill politics, Stabenow sought to leverage all the titles—on specialty crops, organics, biofuels, conservation, rural development, research, education, and nutrition—to build a "pathway" to sixty votes and offset stalemates on the commodity title. The bill she and Roberts introduced on April 23 bore clear evidence of that strategy. On the one hand, the $969 billion package offered $27.6 billion in savings over ten years (even though the bill authorized commodity programs for only five years)— more than the $23 billion promised—by cutting $17.6 billion from a restructured commodity title, $6 billion from underutilized conservation programs, and a little over $4 billion from nutrition. On the other hand, it allocated $2 billion to commodities that lacked baseline funding in the permanent law, including block grants to "specialty crops" of interest to committee members from the Northeast and funds for livestock emergencies sought by members from the Northern Plains. It included elements of the Dairy Security Act being promoted by committee members Leahy, Lugar, and Amy Klobuchar (D-MN) and offered $800 million in spending to promote biofuel development, a pet project of Corn Belt senators. Although the committee's bill ended direct payments *and* the countercyclical target price program agreed to in supercommittee negotiations in favor of the shallow loss insurance program, it tried to mollify southern senators by including targeted crop insurance for peanuts and cotton. The latter was also designed to ward off sanctions by the World Trade Organization (WTO) after its finding that direct payments for cotton violated trade rules.[20] Stabenow maintained the cuts in SNAP generated by raising the LIHEAP threshold to $10 a year but sought to soften the blow by adding $100 million in funding for the Temporary Emergency Food Assistance Program (TEFAP), which provides surplus commodities to food banks. In a bit of symbolic politics, the bill banned SNAP benefits for lottery winners, apparently in response to the story of a Michigan woman who continued to collect food stamps despite winning $1 million in the state lottery.[21]

In short, Stabenow and Roberts navigated strict budget constraints to pull together a coalition they hoped would be sufficient to convince Senate leaders to let the bill go to the floor. Their efforts paid off. Following hours on the phone to their respective colleagues, late-night negotiations among committee staff, and a consequently short two-hour public markup session, on April 26 the committee voted 16–5 to report out S. 3420, the Agriculture Reform, Food, and Jobs Act of 2012.[22] Several senators who voted in favor, notably Roberts, expressed reservations about changes in the commodity title but opted to take their concerns to the floor. Three no votes came from southern Republicans John Boozeman, Saxby Chambliss, and Thad Cochran, who, while praising Stabenow and Roberts for their efforts, complained that the revised commodity title was unfair to their growers. Said Cochran, "On the floor of the Senate, I think we'll have to take advantage of opportunities to offer amendments that may have a chance of strengthening the bill, particularly as it relates to southern interests. Specifically, cotton, rice and peanuts deserve more of a break than they're getting in this bill in this committee."[23] The three also opposed an amendment by Iowa Republican Charles Grassley to impose higher annual revenue thresholds and payment limits on all commodity programs, including insurance. The sole Democrat to vote no was Kirsten Gillibrand of New York, who raised concerns about the impact of SNAP cuts on her constituents and felt that the new dairy program did not do enough to help her state's producers. The final no vote came from Senate minority leader McConnell, who had not attended committee meetings since 2009. As per committee rules, he cast his vote by proxy through Roberts.[24] While McConnell's office offered no explanation for his vote, observers assumed that the Republican leader was siding with his southern colleagues over the commodity title and objected because the bill did not offer the deeper cuts in SNAP spending sought by members of the Senate Republican caucus.

Outside reaction to S. 3420 was as expected. The National Cotton Council thanked committee members for their efforts to align the cotton program with WTO rulings, while the government of Brazil argued that it did not eliminate the trade distortions that had prompted its initial complaint. The Arkansas Farm Bureau and other groups representing peanuts and rice voiced disappointment over the commodity title, while the Specialty Crop Farm Bill Alliance, American Soybean Association, and National Crop Insurance Services offered praise. The National Milk Producers Federation and the National Farmers Union, speaking for smaller milk producers, lauded the new dairy management plan, while the International Dairy Foods Association, representing larger producers and milk processors, panned it. Feeding America, representing the nation's food banks, expressed disappointment in the cuts to SNAP, while the Tea Party–aligned FreedomWorks said the committee had fallen short in cutting "wasteful" federal spending.[25] But they all (with the exception of FreedomWorks) agreed that there *was* a bill to work with.

In the House, Lucas and Peterson praised Stabenow and Roberts but complained that the northern Democrats who dominated the Senate committee had abrogated their earlier agreement by removing the option for countercyclical payments with higher target prices so that they could claim budget savings and then distribute them to other constituencies. Both were adamant that the Farm Bill needed to serve all commodities.[26] "Everybody needs a safety net," Lucas said, but the Senate bill favored Midwest corn and soybean production at the expense of other crops. "If you are in a state that cannot shift into corn and beans," he complained, "the way the language appears to be evolving on the other side of the building, you're in a world of hurt."[27] For his part, Peterson saw the need for a combination of countercyclical payments with higher target price triggers for rice and peanut producers, shallow or catastrophic loss insurance for most other crops, and a new cotton program to meet WTO rules.[28]

Both also knew that any bill coming out of their committee would cut more than the $24.7 billion claimed by Stabenow and colleagues and a lot more than the $4 billion trimmed from SNAP. They had already reacted to Ryan's edict to cut ten-year Farm Bill spending by $30 billion by taking the entire amount out of SNAP, which, they admitted, had largely been a tactical maneuver to comply with a midyear House budget resolution unlikely to be taken up by the Senate. Peterson had even persuaded committee Democrats not to demand a roll-call vote, fearing that if members were forced to go on the record on deep SNAP cuts, it would be harder for them to work together on the overall package. Lucas backed Peterson, vowing that all programs, not just SNAP, would be cut when their committee took up a bill.[29] *When* that would happen, he didn't say—but certainly not until the Senate acted. And at the moment, that wasn't guaranteed.

Reid Fills the Amendment Tree

Nowhere are differences between the House and Senate sharper than on their respective floors. In the House, no bill gets to the floor without first obtaining a "rule" from the Rules Committee stipulating the timing and length of debate, including the number of minutes allotted to each side and, most important, whether the bill will be open or closed to floor amendments. The current era of party cohesion puts the Rules Committee effectively under the control of the Speaker, so surprises on the floor are (or should be) few and far between. Adding to this picture of party control in 2012 was John Boehner's general if unstated adherence to the "Hastert Rule," named after previous Republican Speaker Dennis Hastert (1999–2006), who let no bill go to the floor without the clear support of his party's caucus.[30] The exception to the norm of majority leadership control occurs when 218 of the 435 House members sign a petition to "discharge" a bill out of the authorizing committee or the Rules Committee, sending it directly to the floor for action. Successful discharges are rare; the last one had been in 2002, when a bipartisan coalition bypassed Hastert to discharge out of the Rules Committee what became the Bipartisan Campaign Reform Act (or McCain-Feingold, after its Senate sponsors).

In the Senate, short of obtaining unanimous consent among 100 senators (itself an achievement on anything even remotely controversial), *everything*—whether a bill goes to the floor, when it does so, the length of debate over it, how many amendments are allowed, what it is called—is up for negotiation between the majority and minority leaders and between those leaders and their respective party colleagues. In short, every senator needs—nay, demands—to be consulted. Lurking beneath all this is the ever-present threat of a filibuster by one or more senators, so negotiators on most bills are always on the lookout for sixty votes to cut off debate, a threshold that is even harder to reach when the majority has only a slim advantage, as majority leader Harry Reid had with fifty-two Democrats in 2012.[31]

Making matters even *more* complicated for Senate leaders, and in contrast to the rules of the House, in most cases, amendments need not be germane to the topic of the bill at hand. As a result, the nongermane floor amendment is a favorite way for senators to get pet causes, even entire bills, inserted in otherwise "must-pass" legislation; alternatively, a controversial amendment might be used as a "poison pill" to scuttle a bill outright. No wonder senators tend to select as their party leaders those known for their negotiating skills, their capacity to keep track of members' priorities, and their mastery of the often arcane rules of the chamber.

The personally unassuming Reid had all these traits, and by 2012, he had become particularly vexed at minority leader Mitch McConnell's equally adept use of nongermane amendments paired with motions to suspend the rules even after successful votes to close off debate.[32] Although Reid and McConnell weren't exactly drinking buddies to begin with (overlooking the fact that Reid, a Mormon, didn't drink), relations between the two had become increasingly strained over the past two years as McConnell had used—Reid would say abused—every rule in the Senate handbook to block action on any number of Obama administration proposals, judicial appointments, and even the federal budget. In McConnell's view, he was simply using the chamber's rules and traditions to ensure the minority's right to be considered.

But Reid thought McConnell had gone too far and, backed by members of his caucus, eventually responded with the escalated use of a particularly arcane tactic: filling a bill's "amendment tree" before calling for a cloture vote to end debate—not immediately, but after thirty hours, as prescribed by Senate rules. This tactic, used on occasion by previous majority leaders of both parties, hinged on a combination of Senate rules limiting and giving priority to some types of amendments over others, on the precedent that the majority leader goes first when offering amendments, and, finally, on the majority leader's authority to recognize other senators during floor consideration. As one unsympathetic observer summarized Reid's maneuvering:

> First, he introduces a series of amendments. Next, he offers second-degree amendments that propose minor changes in his original amendments. Under Senate rules, by "filling the tree" and using up the maximum number of amendments allowed, Reid blocks Republican amendments from a vote. Finally, Reid files for a cloture vote, a maneuver designed to end a delaying tactic known as a filibuster. By doing so he forces Republicans to choose: Vote for cloture, which ends floor debate and allows lawmakers to vote on the bill, or vote against cloture, which essentially kills the bill.[33]

Even though McConnell complained loudly and often about Reid's increased use of this tactic, he knew that the rules and norms that promoted the minority party's rights also enabled individual senators to abandon their leaders when push came to shove, especially in an election year. And the Farm Bill, with its bipartisan appeal and array of benefits for specific constituencies, was the kind of package that would result in senators' individual needs trumping party or personal loyalty. Without a ready partisan or ideological divide to rally either side, the real question on S. 3420 was whether Stabenow and Roberts had enough votes to convince Reid and McConnell to go ahead.

They soon had their answer. On May 15 a bipartisan group of forty-four senators sent a letter to Reid and McConnell urging them to bring the bill to the floor.[34] A week later Stabenow announced that she had the sixty votes needed to invoke cloture, even with disputes over the commodity title, and that Reid would give her and Roberts floor time in the first week in June. "We have broad support," she said. "In the Senate these days it seems like you always need 60, but I'm confident that we have 60."[35] She and Roberts thought they could sustain the support of a sufficient number of Republicans, as long as Reid allowed floor amendments. But opening the floor could prove troublesome if senators insisted on nongermane amendments that might turn a nonpartisan Farm Bill into a test of loyalty to party or the president.[36] So the two worked with their respective colleagues to accommodate as many amendments as possible into the "manager's amendment," to be submitted by Reid as the first-order amendment, thus narrowing down the number and range of proposals on the floor.

Of immediate concern was a possible nongermane amendment by Republican John McCain of Arizona requiring the administration to lay out the national security implications of a $500 billion cut in defense spending over ten years should the first phase of the $1.2 trillion sequester take effect in January 2013—prospects for which had increased through 2012 with the stalemate over deficit reduction. Roberts supported McCain in principle but worried that the amendment would insert a partisan poison pill into the delicate coalition supporting S. 3420. "I would prefer it that all non-germane amendments to agriculture be considered in a separate venue, but then again I know that's not possible," Roberts said. "So we are opening the door, like [game-show host] Bob Barker, 'Come on down.' Talk to us. We will try to accommodate anybody and everybody on any section of the bill to try to improve it."[37]

On June 5 Reid moved to consider S. 3420. On June 7 the Senate voted 90–8 on a motion to proceed, with most southern Republicans joining the majority, even though many doubted there was enough time in the current session to get the bill through to the president. Reid then announced that the chamber would take up the bill a week later, giving Stabenow and Roberts time to deal with the 300 amendments already filed. Though Stabenow said she would try to accommodate everyone under her manager's amendment, she would not hesitate to seek Reid's help to move the bill forward.[38] On June 12 Reid moved to fill the amendment tree on S. 3420 to close off nongermane amendments, specifically, one being offered by Republican Rand Paul of Kentucky to cut off US military aid to Pakistan. Reid defended his actions in a floor speech: "I have not given up hope, and I know that Sen. Stabenow has not given up hope, to have a universal agreement so we can really legislate on this bill. . . . We're not going to walk away from this. This bill is far too important, it affects the lives of millions of people."[39]

Reid then applied a litmus test of sorts, using his control over the amendment tree to allow floor action on two germane amendments of considerable policy and political import. The first, on behalf of Democrat Jean Shaheen of New Hampshire, would end the long-standing sugar program, which a wide range of critics considered little more than a market-distorting set of direct and indirect subsidies to politically connected sugar producers. It barely lost by a vote of 50–46, largely because a narrow plurality of senators did not want to upset an already shaky alliance on commodities.[40] The second amendment, on behalf of Senator Paul, would cut nutrition program funding in half and shift control over such programs to the states, a proposal similar to that promoted by Ryan in the House. It lost 65–33, supported only by the most conservative Republicans. Having gauged the temperature on two key issues, Reid then moved on to other matters to enable Stabenow and Roberts to continue their efforts. Over the next several days, the two could be seen negotiating—sometimes loudly—with colleagues on the chamber floor and in the cloakrooms as unrelated business swirled around them.[41]

Floor Action

On June 18 Reid announced a deal with McConnell on S. 3420 that allowed for seventy-three floor amendments, including three nongermane, with the proviso that passage of the nongermane proposals was subject to a sixty-vote threshold.[42] Not included in Stabenow's amendment was any deal on the commodity title, which negotiators had decided to leave to a House-Senate conference, based on the assumption that any House bill would include multiple program options.[43] The proposal to move forward was approved by unanimous consent, testimony to the effort by Stabenow and Roberts to accommodate all comers. Debate and votes on the floor amendments then moved quickly, especially for the Senate.

Successful amendments tended to give modest assistance to particular crops or to make incremental program adjustments. For example, senators voted 58–41 to support an amendment by Maria Cantwell (D-WA) to add more peas, lentils, and chickpeas (major crops in her state) to the federal school lunch program. By voice vote, they approved an amendment by Charles Schumer (D-NY) to enable the USDA to promote the maple syrup industry. Senators also approved amendments by Richard Durbin (D-IL) and Tom Coburn (R-OK) to raise crop insurance premiums for farmers with adjusted gross incomes over $750,000, a change affecting a small number of producers that would save an estimated $100 million a year, and by Saxby Chambliss (R-GA) to require growers getting crop insurance to adhere to federal conservation standards, a move that observers argued was designed less to protect wildlife habitats than to give Chambliss leverage when the time came in conference to bargain over commodity programs.[44] Finally, the body approved by voice vote an amendment by John Kerry (D-MA) on behalf of McCain to repeal a USDA catfish inspection program authorized in the 2008 bill that critics said duplicated Food and Drug Administration food safety procedures and only served to protect domestic catfish producers from Southeast Asian imports.[45]

By contrast, senators defeated amendments that might have upset previous deals or that had little chance in the House or with the president. They voted down (56–43) an effort by Jeff Sessions (R-AL) to sharply reduce SNAP eligibility, even as they opposed (66–33) an amendment by Kirsten Gillibrand (D-NY) to cut subsidies to companies selling crop insurance to offset the now $4.5 billion cut in SNAP. Efforts by the Senate's more libertarian Republicans to eliminate biofuel subsidies and slash a range of commodity programs met with general bipartisan disapproval, with most senators being content to maintain the status quo so long as the overall bill hewed to its promised budget savings. Also going down to defeat (73–26) was the last germane amendment, by Independent Bernie Sanders of Vermont, to permit states to enact laws mandating labels on foods containing genetically modified ingredients. Senators then disposed of the three nongermane amendments, voting 95–4 on an amendment by Tom Coburn (R-OK) to prohibit the use of public funds for political party conventions; by voice vote in support of an amendment by Patty Murray (D-WA) to require the administration to submit detailed reports of the *overall* (not just defense) effects of the sequester; and 54–45 against a proposal by Marco Rubio (R-FL) that critics said would weaken the authority of the National Labor Relations Board to enforce union contracts with employers.

The final vote on S. 3420, the Agriculture Reform, Food, and Jobs Act, was 64–35. All but five no votes came from Republicans, either because they disapproved of the bill's spending or, in the case of Thad Cochran (Mississippi) and Richard Shelby (Alabama), because they were unhappy with the commodity title's treatment of southern crops. Democrats who voted no either were from rice-growing states (David Pryor of Arkansas and Mary Landrieu of Louisiana) or opposed the cuts in SNAP spending (Frank Lautenberg of New Jersey and Jack Reed and Sheldon Whitehouse of Rhode Island). Of note, Gillibrand voted yes, to support her committee chair and keep her options open.[46]

Onward?

In passing S. 3420, the Senate acted with what observers regarded as uncharacteristic speed. "It is 2 o'clock in the afternoon, not 2 o'clock in the morning," exulted Reid, while McConnell called the vote "one of the finest moments in the Senate in recent times in terms of passing a bill." They credited Stabenow and Roberts, who in turn praised their colleagues. "This bill represents significant reform," said Stabenow. "It cuts subsidies, it cuts the deficit and it creates jobs." Roberts called the legislation the best bill possible: "It shows what can happen if we break the logjam of partisanship and work together to get something done."[47]

Both senators expressed hope that their colleagues in the House Committee on Agriculture would be able to pick up where they left off. "I call on the leadership in the House to work with them as the Senate leadership worked with us," Stabenow said. For his part, House Agriculture Committee chair Frank Lucas announced that his committee would start work on July 11. "Although there will be differences between the Senate approach and our own," he said, "I hope my colleagues are encouraged by this success when we meet on the 11th to consider our own legislation."[48]

7

We're on the Road to Nowhere

A lot of times they don't pass a lot of stuff on the House side anymore, because they don't have the rural votes for it. . . . There wouldn't be a farm policy right now or in the future weren't it for the Senate.
—Southern senator

Whatever Frank Lucas hoped might happen as he convened the House Agriculture Committee on July 11, 2012, even he may have been wondering whether the House would act before the end of the 112th Congress, much less before the 2008 law expired on September 30. Indeed, many senators had supported S. 3420 knowing that, so far in 2011–2012, they had voted in bipartisan majorities to pass a transportation bill, post office reform, and a reauthorization of the Violence against Women Act, only to see no action whatsoever on these measures in the House.[1] What were the odds that the House would take up the Farm Bill before the fall election, or even after it?

Such doubts were fed by the ongoing stalemate over the budget, particularly the extent of immediate and long-term cuts to nondefense discretionary spending. In April Speaker John Boehner and majority leader Eric Cantor had backed a proposal by Budget Committee chair Paul Ryan to cut an additional $19 billion in nondefense appropriations for the remainder of fiscal year 2012, effectively canceling the previous summer's agreement with President Obama. That action, supported by a majority of the House Republican caucus, prompted threats of a presidential veto and raised prospects of a government shutdown at end of the fiscal year—a month before the November elections. It would also complicate House action on the Farm Bill, due to expire at the same time. Most first-term conservative Republicans, largely hailing from safe Republican districts, wanted even deeper cuts in nondefense spending and were not likely to support a bill calling for nearly $1 billion in outlays over ten years. Some of their longer-serving colleagues, concerned about challenges from their right, wanted to avoid being portrayed as big spenders; others worried that any floor fight over cuts in nutrition programs would give the Democrats an issue to raise in the general election.

As a result, some members and outside observers were already downplaying any doomsday scenarios of not reauthorizing by September 30. For one thing, on major commodities like corn and soy, the practical effects of delay would not be felt until the spring 2013 planting season, so producers might be able to live with uncertainty for a few more months. Moreover, delays were not that unusual—they had happened in 1996 and 2008—and Congress could always pass a short-term extension to ensure continuity in programs (such as dairy) in which a lapse of legal authorization mattered.

But Lucas would have none of it. Though he admitted that a short-term extension was a possibility, "I can't have my producers, my fellow farmers and ranchers, caught in a twilight zone between the old farm bill and the new farm bill," he said.[2] For his part, ranking minority member Collin Peterson thought he and Lucas had the votes to get a bill out of committee. "The bigger problem," he admitted, "is going to be, first of all, getting it onto the floor, getting leadership to give us time. And even then it's going to be tough. The left won't be satisfied, the right won't be satisfied and there aren't that many people in the middle anymore."[3]

The Center Holds, for Now

But there *was* a center of sorts on the Agriculture Committee, albeit a small and shaky one anchored by Lucas and Peterson and extending to a comparative minority of rural legislators in both parties. Their goal was to get a bill to a conference committee as soon as possible—ideally, one that did not stray too far from the package worked out with Stabenow and Roberts the previous fall.

On July 9 Lucas and Peterson formally filed H.R. 6083, the Federal Agriculture Reform and Risk Management Act, or FARRM. Aside from their titles, the contrasts between H.R. 6083 and S. 3420 were apparent but not particularly insurmountable. The former called for $35 billion in cuts over ten years, compared with the $23.6 billion in the Senate bill, largely by taking another $16 billion away from SNAP. As Lucas and Peterson promised, it also offered growers a choice between "shallow loss" insurance coverage and price loss coverage triggered by higher target prices, paying for the latter with the additional cuts in nutrition programs.[4] Lucas knew that various committee members wanted other major changes, but he asked them to save any particularly controversial amendments for the House floor. "If you want to dramatically reshape entire programs, if you want to do things of [an] unimaginable bold nature, that's probably floor work," he said.[5]

Committee members listened, for the most part. On July 11, following a late-night session, the committee voted 35–11 to report out H.R. 6083 with comparatively few changes. Although the center more or less held, the amendments debated in committee reflected deep divisions over commodity and nutrition programs. One involved dairy: members voted 29–17 against an amendment by Bob Goodlatte (R-VA) and David Scott (D-GA) to replace the dairy stabilization program favored by smaller dairy producers and adopted by the Senate in S. 3420 with an insurance program that was more acceptable to larger dairy operations and milk processors. Lucas and Peterson—coauthor of the dairy stabilization program—both voted no. Another Goodlatte amendment, filed at the behest of candy makers, would have eliminated the heavily criticized sugar program. Like Shaheen's floor amendment in the Senate, it too went down to defeat, with Lucas and Peterson arguing against major changes in programs affecting specific crops. On commodity programs, at least, there was a good chance that H.R. 6083 and S. 3420 could be reconciled in conference, assuming there were no major changes on the House floor.

More telling were proposed amendments on nutrition programs. Although the committee's Republican majority approved provisions to tighten SNAP eligibility requirements, Lucas and a few other Republicans sided with committee Democrats in rebuffing even more drastic changes proposed by the most fiscally conservative newcomers, including an amendment by Tim Huelskamp of Kansas to double SNAP cuts to $33 billion.[6] For his part, Peterson sided with Lucas and the Republicans in defeating amendments by the committee's more liberal Democrats—a significant bloc of the minority's twenty members—to eliminate cuts in nutrition programs entirely or limit them to the $4.6 billion called for in S. 3420. Not all votes on nutrition programs were so divisive: the committee approved by voice vote amendments by Chellie Pingree (D-ME) and Tim Johnson (R-IL) to make it easier to use SNAP benefits to purchase healthy fruits and vegetables.[7]

Four committee Republicans voted no on H.R. 6083. Three of them—Huelskamp, Marlin Stutzman (Illinois), and Joseph Gibbs (Ohio)—were Tea Party affiliates for whom the bill failed to rein in what they considered out-of-control SNAP spending. Huelskamp made it clear that conservatives even might offer a floor amendment to split nutrition and farm programs into separate bills. "I think we need some assurances on the floor that we can get some food stamp reforms," he said. "There's going to be a fight on that. I'm not afraid to fight."[8] Goodlatte of Virginia voted no largely over the dairy and sugar programs. The seven Democrats opposed were New Englanders Pingree (Maine), Joseph Courtney (Connecticut), and James McGovern (Massachusetts); Joseph Baca of California; and three African American members—Marcia Fudge (Ohio), Scott (Georgia), and Terri Sewell (Alabama)—because of the $16.5 billion in SNAP cuts. In a harsh dissent to the committee bill, they pointed to a Congressional Budget Office estimate that the cuts would cause 2 million to 3 million people to lose SNAP benefits and nearly 300,000 children to become ineligible for free school lunches. "These cuts are immoral," they declared. "They are hurtful. And they are exactly the wrong answer for people who struggle with hunger."[9]

Outside reactions were instructive. While farm and commodity groups expressed general support for H.R. 6083 as a step toward reauthorization, conservative advocacy groups excoriated Lucas and the committee's majority for fiscal profligacy. "I don't think the leadership and many of the freshman Republicans who ran on cutting spending want to pass a nearly trillion-dollar bill this close to an election," said Steve Ellis, vice president of Taxpayers for Common Sense.[10] Another conservative group, the Madison Project, was especially blunt:

> Last week, the House Agriculture Committee marked up the preposterous $957 billion farm/food stamp bill (H.R. 6083). Despite the media reports about severe

cuts to the food stamp program, this bill actually locks in the appalling levels of spending established in the Obama-era. In many respects, the committee votes from last week shed light on the problem we have with many red state statists within the Republican Party.[11]

On the other side of the ideological spectrum, nutrition and hunger advocates echoed committee Democrats in condemning the $16 billion in nutrition program cuts. Many began to express a preference for a simple one-year extension of the 2008 law, with no cuts in SNAP, rather than trying to pass a five-year reauthorization under the current political circumstances.[12]

Debbie Stabenow applauded Lucas and Peterson for their achievement and urged House leaders to bring the bill to the floor to keep Congress on track to pass a bill before September 30: “Letting current farm policy expire and reverting back to the policy of the 1940s or kicking the can down the road with a short-term extension will not only hurt our economy, it will also mean a lost opportunity to enact major reforms in farm policy and substantial spending cuts.”[13] H.R. 6083 went to the Rules Committee on July 24, but when—if—it would act was anyone’s guess. Peterson worried about delaying action until after the fall elections, when any bill might be caught up in the still unresolved battle over the federal budget and debt limit. Such a fight would be a “quagmire,” he warned, raising the probability of the Farm Bill being “taken away from us.”[14]

Dairy Wars, Food Stamps, and Drought

Lucas and Peterson had been able to thread the proverbial needle in committee, but once H.R. 6083 was reported out, action on it stalled almost immediately, held up by sharp disagreements within the House Republican caucus and by leadership concerns about the potential electoral ramifications of a divisive floor fight.

One critical intraparty division quickly emerged between Lucas and Boehner over the dairy price support programs in H.R. 6083 and S. 3420. Boehner, a longtime skeptic of commodity supports and an ally of Kroger Foods, a major supermarket chain based near his southern Ohio district, opposed dairy “management,” calling it contrary to free markets and little more than “Soviet-style” production control. He had backed Goodlatte’s amendment to eliminate Peterson's dairy program, which he was determined to keep out of a conference committee, where it might get locked into whatever consensus package emerged. Lucas, always seeking to support as many farm constituencies as possible, worried that losing a dairy program favored by northern and upper midwestern members of both parties would undermine the bipartisan coalition needed for Farm Bill passage.

A more critical intraparty cleavage emerged over SNAP. Republicans outside the Agriculture Committee, led by majority leader Cantor and Budget Committee chair (and soon to be vice presidential nominee) Ryan, thought that Lucas and his committee colleagues had not gone far enough in cutting overall spending on nutrition programs in particular. Ryan had earlier proposed to move SNAP to a stand-alone block grant to the states, resulting in (depending on scenarios) anywhere from $30 billion to $134 billion in potential savings over ten years.[15] Cantor, whose desire for Boehner's job was little disguised, was backed by the pivotal bloc of first-term conservatives, several of whom promised floor amendments that would succeed where the committee had failed. Although Lucas agreed in general on the need to "reform" SNAP, he opposed separating it from farm programs. Doing so would only alienate most Democrats, for whom the longtime "farm programs + food stamps" deal had become their only reason to support commodity subsidies.

Complicating matters for Lucas was a lack of urgency among commodity producers to counter the zeal of the fiscal purists in his party. General farm organizations urged full reauthorization to avoid lapses in program coverage, but out in the fields, corn and soy growers in particular were enjoying record prices and revenues. Even those being affected by an ongoing drought in the upper Midwest and Plains had crop insurance sufficient to cover their losses. Absent concerted grassroots pressure from key rural constituencies, Boehner could let the Rules Committee sit on H.R. 6083 indefinitely.

As a result, any hope of moving H.R. 6083 that summer depended on the ability of a small group of rural Republicans in competitive seats to convince the leadership to let the bill go to the floor. The ongoing drought—the worst some areas had seen in five decades—was a particular problem for livestock producers in the upper Plains, and the situation became more acute after a special livestock disaster program in the 2008 bill expired at the end of fiscal year 2011. Their concerns led first-term Republican Kristi Noem of South

Dakota to issue a letter urging Boehner and Cantor to let H.R. 6083 go to the floor by the end of July.[16] Cantor was not persuaded. "They said they're concerned the bill wouldn't pass," Noem explained. "They obviously don't want to send the farm bill to the floor to have it fail." Boehner even mentioned the drought when asked about the bill. "Most farmers in my district . . . avail themselves of crop insurance," he said. "That's why it's in the farm bill, that's why our government subsidizes the cost of crop insurance, to encourage farmers to buy that. In most cases, it should be sufficient to deal with this drought."[17]

Lucas soon turned to another option. In late July he proposed a one-year extension of the 2008 law to include targeted disaster aid for livestock, fruits, and vegetables. Boehner initially expressed support for the idea, which would avoid an election-year floor fight over food stamps, but he quickly pulled back when it became clear that it faced opposition in the Republican caucus.[18] A week later, on August 2, the House passed a disaster assistance bill directed at livestock producers and intended to help members like Noem who were engaged in tough reelection battles. The relief package, funded by cuts in conservation programs, passed 223–197, largely along party lines. Of note, forty-six Republicans, most of them fiscal conservatives, voted no. They opposed any new spending. However, their votes were offset by the support of thirty-five Democrats, most of them from rural districts in the Midwest and South. They included Peterson, who nonetheless called the bill "a sad substitute for what is really needed—long-term farm policy." Farm groups were similarly underwhelmed, and few expected the measure to be considered in the Senate before the August recess, if at all, since disaster relief was already included in S. 3420.[19]

Congress soon went on August recess, leaving prospects for a bill before September 30 poorer than ever. The hiatus also gave Farm Belt Democrats an opportunity to hammer Republicans on House inaction. Noem of South Dakota and Rick Berg of North Dakota took heat from their opponents for their inability to get Boehner and Cantor to move on the Farm Bill, even as the region's livestock industry suffered.[20] At the same time, conservative advocacy groups did everything possible to shore up *their* Republicans, arguing that no bill at all was better than the one on the table. For Huelskamp of Kansas, the issue that resonated with his constituents was not the lack of a Farm Bill per se but the out-of-control spending on SNAP. "For conservatives like myself, the real concern is that what is now a farm bill is really not that. *It's a food stamp bill*," he said. "When Obama says that we will cut not a dime from the food stamp program, which has increased 72 percent in spending in three and a half years, you can't be serious."[21]

A coalition of more than fifty farm and commodity groups, putting aside their own economic and ideological differences, tried to rally Congress into action when it reconvened after Labor Day. Notable in their absence from the lobbying effort were nutrition and hunger groups, for whom the SNAP cuts in the House bill were too much. Knowing that food stamps were exempt from the budget sequester, they had decided to sit this one out and hope for action on the Senate bill or make do with no new bill until 2013. On September 13 Democrat Bruce Brailey of Iowa filed a formal petition to discharge H.R. 6083 from the Rules Committee. Although the effort drew 66 signatures from members of both parties, including a few Republicans in tight election races (including Noem and Berg), it was not supported by Lucas and Peterson. Both of them were unwilling to fracture their relationship with the leadership, and the petition ultimately fell far short of the 218 signatures needed.

Prospects for House consideration of H.R. 6083 before September 30—and, in effect, before the November 8 election—died in mid-September when Lucas could not get the Republican caucus to support a three-month extension of the 2008 law.[22] Most Democrats wanted full reauthorization and opposed the stopgap measure, while a core group of conservative Republicans wanted even the three-month extension to require the $39 billion in cuts included in H.R. 6083. Republican leaders on both sides of Capitol Hill also wanted to avoid a public fight over nutrition programs right before the general election. They were hoping that a Mitt Romney victory, and possibly a Republican takeover of the Senate, would change the political equation in 2013. Senate Democrats, for their part, were furious that the House had once again failed to act on legislation their chamber had passed with a bipartisan majority. They decided that allowing the 2008 law to expire might spur House leaders to act on their bill after the election and give a conference committee the opportunity to work out a compromise before the end of the 112th Congress. "We're at a stage now where it's been a total failure . . . of leadership in the House," said majority leader Reid. "Just to walk away from this? And that's what they're doing."[23]

In the absence of any effective grassroots pressure, and with House leaders confident that the full impact of a failure to reauthorize would not be felt for months, Congress let the 2008 law expire on October 1. It was the first time in history that the House had failed to act on a Farm Bill reported out of the Agriculture Committee as its deadline passed.[24] Ironically (to some), Congress managed to pass a continuing resolution on the budget through March 2013, keeping the amount spent on nutrition programs at current levels.[25] Collin Peterson summed up the feelings of many colleagues: "No wonder no one likes Congress. Members will now have to explain to their constituents why the House did not even try to consider a new five-year farm bill."[26]

Elections Have (Some) Consequences

Agricultural issues barely merited discussion during the 2012 presidential election campaign, likely because neither major-party candidate thought rural votes were up for grabs. In the end, rural America went for Mitt Romney, with 61 percent of rural voters supporting the Republican candidate, and rural states in the Midwest and South accounted for most of his 206 electoral votes. More telling, in one preelection poll, 75 percent of farmers producing on 500 or more acres said they would vote for the Republican.[27] But as we know, Barack Obama won reelection rather handily, getting the bulk of his 332 electoral votes from the more populated states on both coasts and in the Great Lakes region. Reflecting broader demographic, ideological, and partisan splits, Obama's strongest support, even in the states he lost, came from voters living in cities with populations of 50,000 and larger. The implications of the election on the president's future stance on agricultural policy were not yet clear, but observers weren't betting on his enthusiasm for commodity programs.

Equally important, Senate Democrats managed to increase their majority by two seats, to fifty-three. A loss in Nebraska was offset by gains in Maine, Connecticut, and, notably, Indiana. Democrats held on to seats in North Dakota, Missouri, and Montana that many observers thought might go to Republicans.[28] Debbie Stabenow, no doubt aided by a strong Obama victory in Michigan, won easier reelection than many had predicted. In North Dakota, Congress's failure to act was a factor in state attorney general Heidi Heitkamp's victory over Representative Rick Berg for the seat vacated by the retiring Kent Conrad. Heitkamp had heaped scorn on Berg for failing to get more House Republicans to sign the discharge petition.[29]

In the House, Republicans suffered a net loss of eight seats but kept their majority. Mirroring the presidential race, and of particular interest to the future of Farm Bill policymaking, House elections in 2012 resulted in Republican gains in the rural Midwest and South, with districts in Arkansas, Indiana, Kentucky, North Carolina, and Oklahoma shifting from moderate or conservative Democrats to even more conservative Republicans. Republican losses tended to be in coastal states such as California, Florida, New Hampshire, and New York and in the Great Lakes states of Illinois and Minnesota. Of note, ten of the seventeen Republicans who lost reelection were conservatives first elected in 2010, largely in competitive suburban districts.

How the 2012 elections would affect the House Republican caucus in 2013 was yet to be seen, but the immediate impact seemed to strengthen Speaker Boehner's hand—or at least his resolve—with respect to his most conservative critics as he approached negotiations over avoiding the budget sequester set for January 2013. On December 4 the House leadership dropped four members from key committee posts as punishment for their persistent lack of loyalty on key votes. Notable among them was Tim Huelskamp, ousted from the Budget Committee and, just to rub it in, the Agriculture Committee as well. "This is clearly a vindictive move," Huelskamp responded, "a sure sign that the GOP establishment cannot handle disagreement."[30]

Conservative activists cheered, with the Club for Growth noting that Huelskamp and colleagues were now "free from the last remnants of establishment leverage against them."[31] But agricultural groups back in Kansas were stunned. "It certainly puts our members and Kansas as one of the top ag states at a significant disadvantage in setting federal policy," said Aaron Popelka of the Kansas Livestock Association. Justin Gilpin of the Kansas Wheat Commission said, "Having somebody from the Big First on House Ag has been extremely important and deserving because of the amount of wheat and agriculture that come from there."[32] Seasoned observers of Huelskamp's career were not surprised. "He's earned it," said noted Kansas State University agricultural economist Barry Flinchbaugh, not known for mincing words. "You represent the largest single agricultural district in the country. You don't go voting against farm bills in this kind of a situation. You don't make food stamps the issue. You certainly don't vote in favor of taking food stamps out of the farm bill when everyone else—including myself—told him it would be a disaster."[33]

Despite such warnings, Huelskamp had just been reelected, without opposition.

The Aggies Lose Control

When the 112th Congress reconvened after the November elections, members faced a stack of unfinished business, the Farm Bill included. The top priority, however, was to avoid the so-called fiscal cliff and cut yet another deal on the debt limit and budget before January 1, 2013, when the stopgap measures enacted during the 112th Congress would expire. Without action, Bush-era tax cuts for millions of Americans would expire; the government would, once again, technically default on its debts; and automatic sequestration on all federal discretionary spending would kick in. In theory, nobody wanted to go there. However, the elections had done little to alter the fundamental balance of power in Washington, so the core question was: who would make what concessions on taxes and spending?

As in December 2011, negotiations over any budget deal would take place among a small group of House and Senate leaders and, eventually, between them and President Obama. It soon became clear that fiscal-cliff negotiations would take priority over all other business in Congress, effectively eliminating any prospect for action on most stand-alone bills. This was particularly true in the House, where Lucas could not get Boehner to bring H.R. 6083 to the floor, despite public pleas by farm and commodity groups to pass a bill in the 112th Congress.

For commodity interests, and for dairy producers in particular, the pressing question was whether any elements of S. 3420, H.R. 6083, or the already expired 2008 law could be included in a budget deal. A second question was whether they, or even House and Senate Agriculture Committee leaders, would have much say in it all. These were not mere theoretical concerns. Although the expiration of the 2008 law on September 30 would not affect most major commodity producers until spring planting season, its lapse would soon have consequences for many farm operations, especially smaller ones that depended on loan, grant, and conservation programs that were not part of the permanent law. Even more important, it would soon affect bulk—and hence consumer—milk prices: once the ongoing dairy price support program formally expired on January 1, the USDA would be required to pay 1949-era prices for bulk milk, which, according to experts, would *double* consumer dairy prices within months if not sooner.[34] To most consumers, the prospect of a "milk crisis" was the only tangible result of a failure to renew the Farm Bill, and it got headlines. Otherwise, few paid attention. As Secretary of Agriculture Thomas Vilsack summed up the situation in December 2012: "Why is it we don't have a farm bill? It isn't just differences of policy, it is because rural America with its shrinking population is becoming less and less relevant to the politics of the country."[35]

Early negotiations on the fiscal cliff offered some hope for a stand-alone Farm Bill, with the Obama administration proposing cuts in commodity, conservation, and insurance programs as part of an overall budget deal that would reduce USDA spending by $32 billion over ten years, compared with the $23 billion in S. 3420 and the $35 billion in H.R. 6083. The administration tellingly proposed *no* cuts in SNAP, an opening gambit for negotiations with Boehner and Senate minority leader McConnell—but not the chairs of the House and Senate Agriculture Committees—since failure to get a deal on the budget made everything else moot.[36] Whatever Boehner's willingness to cut such a deal, it soon became clear that House conservatives, shored up by their outside allies, were still adamant about getting deeper cuts in SNAP than Obama would ever accept.[37] Any chance that H.R. 6083 would get to the House floor, or that Boehner would allow the House to vote on S. 3420 as a substitute, faded.

Faced with that reality, House and Senate Agriculture Committee leaders tried to hammer out an agreement on commodity and nutrition programs that could be included in any fiscal-cliff deal, but their negotiations broke down largely over the House's failure to act on H.R. 6083. When Lucas accused the Senate committee of being unwilling to negotiate, Stabenow was blunt about the position taken by her colleagues. "We have passed a bill through the entire body," she said. "They haven't passed it through the House. There has to be leadership in the House to care about rural America and to back up their leaders." Secretary Vilsack, meanwhile, warned Agriculture Committee leaders of the consequences of failing to reach an agreement. "The risk the Agriculture leaders face is that the speaker and the president could get in a room," Vilsack said, suggesting that those two could take whatever they wanted out of agriculture in making their own deal on the budget.[38]

Faced with such a prospect, in late December the Agriculture Committees settled for a simple extension of the 2008 law to September 30, 2013. But by that time, just as Peterson had feared months earlier, they had lost control. Boehner, stymied by hard resistance within his ranks over allowing any of the Bush tax cuts to lapse, essentially gave up and handed over the Republican negotiating reins to McConnell, who made it clear that he might extend only some portions of the 2008 law as part of any deal. The irony that McConnell now had control over negotiations, despite years of an apparent lack of interest, was not lost on anyone. "McConnell doesn't understand agriculture policy, yet he's trying to write a new farm bill extension in the fiscal cliff deal, rather than take a bipartisan recommendation from leaders of the House and Senate Agriculture committees," said one committee aide.[39] A furious Stabenow was blunter: "I want to hear somebody justify that on the floor," she said. "People are sitting in rooms trying to decide how we get deficit reduction, and we passed something that saves $24 billion in a fiscally responsible way. . . . We went through every single page of the farm bill, which is what we ought to be doing in every part of government. . . . We did that. And now, at the last minute, none of that matters?"[40]

The House and Senate Agriculture Committees rushed to put together a compromise extension that McConnell might support. But McConnell and Boehner had already agreed on one thing: no extension of any length would include the dairy management program contained in the Senate and House bills. For Boehner, the impending "milk crisis" resulting from expiration of the 2008 law was no reason to implement a program he openly assailed as "Soviet-style communism." He still refused to budge as late as December 30, when he argued with Lucas in front of the Republican caucus over proposals to extend the 2008 law. Boehner and Cantor backed short-term extensions—one of them for only thirty days, which Peterson called "a poor joke on farmers"—while Lucas and fellow House Agriculture Committee members supported a simple nine-month extension.[41]

The tax and budget deal cut between McConnell and Vice President Biden on December 31 included provisions that extended the 2008 law through September 2013. Although the package McConnell put together adjusted dairy programs to prevent an increase in milk prices in January, it pointedly left out the dairy management program supported by both Agriculture Committees. Equally telling, McConnell left intact commodity programs with baseline funding—notably, the much-criticized direct payment system—to the evident satisfaction of Roberts and southern Republicans. As if to needle Stabenow directly, he paid for those commodity supports by failing to fund programs whose authorization had lapsed at the end of fiscal year 2012—notably, programs for conservation, organic farming, and fruit and vegetable production. Nor did McConnell's package contain the disaster assistance sought by Plains states Democrats.[42]

If McConnell had been brazenly political in putting together the agriculture portions of the tax and spending deal, he did so knowing that he needed the votes of his Republican colleagues, particularly those in the South. Therefore, he was willing to sacrifice many of the comparatively small deals worked out in passing S. 3420 for the sake of preserving the range of commodity programs popular with his core party base. He also knew that a newly reelected President Obama had the muscle to enforce any deal among otherwise disappointed Farm Belt Democrats. Moreover, Obama had made it clear from the start that any deal could not touch SNAP spending, a far greater priority for his party's urban base than any of the Farm Bill titles left underfunded.

Stabenow, stung by McConnell's dismissal of the compromises made in S. 3420, aimed her criticism at the minority leader: "Rather than embrace the Senate's bipartisan farm bill, which cuts $24 billion in spending and creates certainty for our agriculture economy, Senator McConnell insisted on a partial extension that reforms nothing, provides no deficit reductions, and hurts many areas of our agriculture economy."[43] Peterson was furious at McConnell's last-minute substitution of his own package for anything proposed by the Agriculture Committees, and with Cantor for refusing to let H.R. 6083 go to the floor, but he was just as angry with his own party's leaders. He accused Obama and Biden, with the tacit approval of majority leader Reid, of selling out Stabenow and the Agriculture Committees for the sake of a deal—so long as consumer milk prices didn't go up. "Upset is an understatement," Peterson fumed. "I'm not going to talk with those guys. I'm done with them for the next four years. They are on their own. They don't give a sh-it about me, anyway."[44]

On January 1, 2013—hours after the nation had technically gone over the "fiscal cliff"—Congress formally approved the American Taxpayer Relief Act of 2012 (H.R. 8), which, as its name suggests, largely focused on restricting to higher earners the impact of the lapsing Bush tax cuts. In doing so, as conservatives would bitterly complain, it still effectively raised billions in new federal revenue for future years. On the spending side, however, the deal kicked the proverbial can down the road by extending the debt limit and sequestration deadlines to March 2013. Let the next Congress deal with it.

The Senate voted 89–8 to approve H.R. 8. In the House, conservative Republicans, led by majority leader Cantor, refused to back a deal that included revenue increases of any kind. But Boehner, breaking with his normal adherence to the Hastert Rule, brought H.R. 8 to the floor anyway. It passed by a vote of 257–167, with 172 Democrats in favor and 151 Republicans, including Cantor, opposed. Peterson, angry about how the deal treated agriculture, was a rare Democratic no vote. Observers also noted that Boehner, ignoring the custom of the Speaker voting only to break a tie, cast his vote with the majority.[45]

Thus ended the 112th Congress. The 113th would convene two days later.

Figure 8.1 Timeline of Farm Bill, 113th Congress

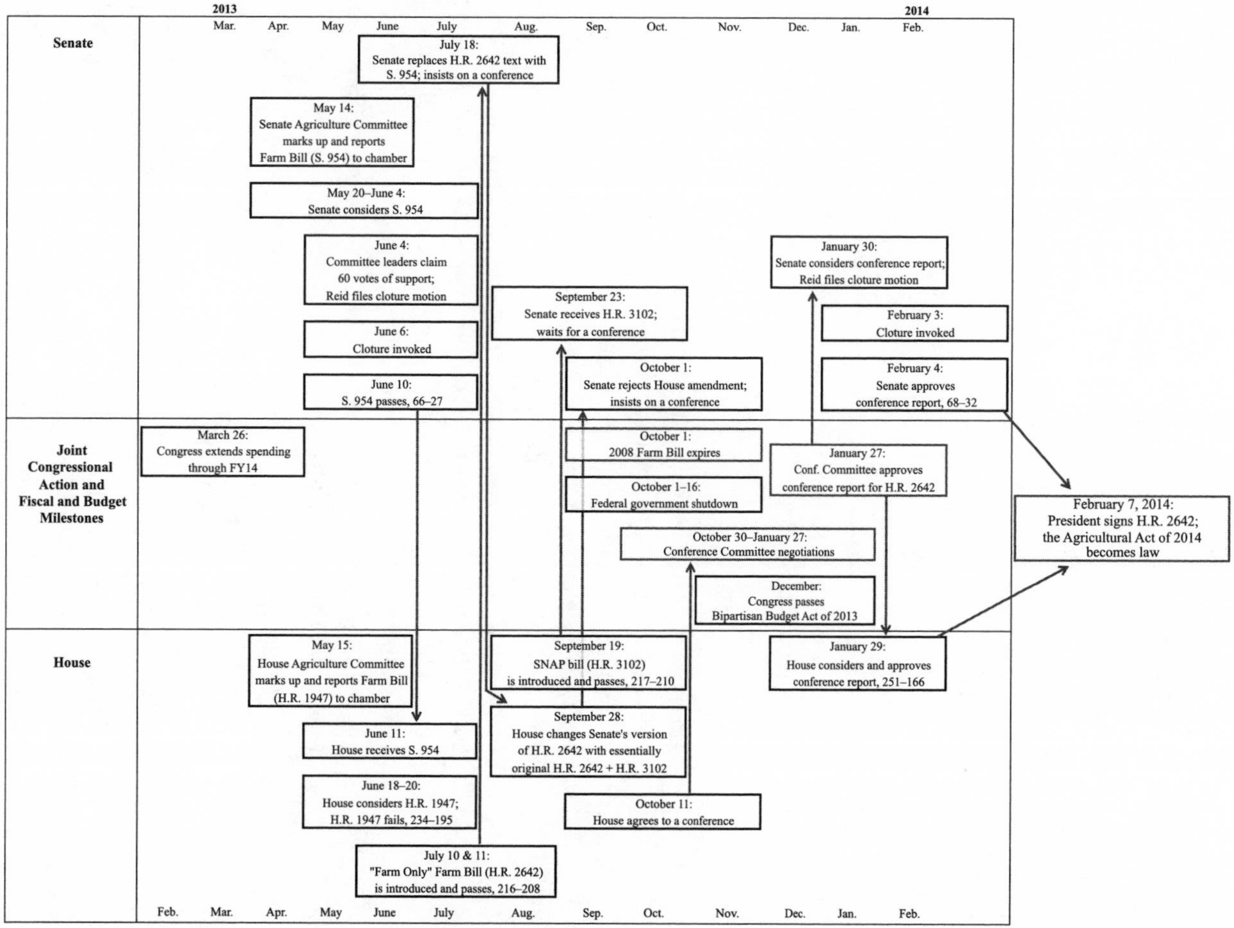

2013 — 2014

Mar. Apr. May June July Aug. Sep. Oct. Nov. Dec. Jan. Feb.

Senate

July 18:
Senate replaces H.R. 2642 text with S. 954; insists on a conference

May 14:
Senate Agriculture Committee marks up and reports Farm Bill (S. 954) to chamber

May 20–June 4:
Senate considers S. 954

June 4:
Committee leaders claim 60 votes of support;
Reid files cloture motion

June 6:
Cloture invoked

June 10:
S. 954 passes, 66–27

September 23:
Senate receives H.R. 3102;
waits for a conference

October 1:
Senate rejects House amendment;
insists on a conference

January 30:
Senate considers conference report;
Reid files cloture motion

February 3:
Cloture invoked

February 4:
Senate approves conference report, 68–32

Joint Congressional Action and Fiscal and Budget Milestones

March 26:
Congress extends spending through FY14

October 1:
2008 Farm Bill expires

October 1–16:
Federal government shutdown

October 30–January 27:
Conference Committee negotiations

December:
Congress passes Bipartisan Budget Act of 2013

January 27:
Conf. Committee approves conference report for H.R. 2642

February 7, 2014:
President signs H.R. 2642;
the Agricultural Act of 2014 becomes law

House

May 15:
House Agriculture Committee marks up and reports Farm Bill (H.R. 1947) to chamber

June 11:
House receives S. 954

June 18–20:
House considers H.R. 1947;
H.R. 1947 fails, 234–195

July 10 & 11:
"Farm Only" Farm Bill (H.R. 2642) is introduced and passes, 216–208

September 19:
SNAP bill (H.R. 3102) is introduced and passes, 217–210

September 28:
House changes Senate's version of H.R. 2642 with essentially original H.R. 2642 + H.R. 3102

October 11:
House agrees to a conference

January 29:
House considers and approves conference report, 251–166

Feb. Mar. Apr. May June July Aug. Sep. Oct. Nov. Dec. Jan. Feb.

8

SNAP

Any farm bill representing the real views of the agrarian Southern and Midwestern members can pass only when most urban members are absent.
—Former undersecretary of agriculture John A. Schnittker

On the surface, the 113th Congress that convened on January 3, 2013, did not look too different from the 112th that had adjourned *sine die* two days earlier. Although House Republicans held eight fewer seats—largely due to losses by first-term conservatives in suburban northern districts that had voted for Obama over Romney in the presidential race—they suffered comparatively little and started 2013 with a 231–204 majority. John Boehner was reelected Speaker, but not before overcoming an effort by a dozen disgruntled conservatives, including Tim Huelksamp, to rally support for one of their own.[1] In the Senate, Democrats picked up two seats, raising their total to fifty-three—or fifty-five when Independents Angus King of Maine and Bernie Sanders of Vermont were added to the Democratic caucus—giving majority leader Reid a bit more leverage in his duels with minority leader McConnell.[2]

The relative absence of major overall change was apparent on the Senate Agriculture Committee (table 8.1). There were two new Democrats: Heidi Heitkamp of North Dakota took the seat vacated by retiring fellow Democrat Kent Conrad, and Joe Donnelly of Indiana replaced veteran Republican Richard Lugar, who had been defeated in his primary by a Tea Party–affiliated state treasurer whom

Table 8.1 Members of the Senate Committee on Agriculture, Forestry, and Nutrition, 113th Congress

	State	Year First Elected
Democrats (11)		
Debbie Stabenow (chair)	MI	2001
Patrick Leahy	VT	1974
Tom Harkin	IA	1984
Max Baucus	MT	1978
Sharrod Brown	OH	2006
Amy Klobuchar	MN	2006
Michael Bennet	CO	2009
Kirsten Gillibrand	NY	2008
Robert Casey Jr.	PA	2006
Joe Donnelly	IN	2012
Heidi Heitkamp	ND	2012
Republicans (9)		
Thad Cochran (ranking minority)	MS	1978
Mitch McConnell	KY	1984
Pat Roberts	KS	1996
Saxby Chambliss	GA	2002
John Boozeman	AR	2010
John Hoeven	ND	2010
Mike Johanns	NE	2008
Charles Grassley	IA	1980
John Thune	SD	2004

Donnelly then beat in the general election. Reflecting the slightly enhanced Democratic majority in the Senate, Donnelly kept Lugar's committee seat, raising Debbie Stabenow's majority by one. More important, Republican Thad Cochran of Mississippi invoked seniority to bump Pat Roberts as the committee's ranking minority member. Observers expected Cochran to be more forceful in pushing Stabenow and her fellow northern Democrats to revisit the commodity title on terms friendlier to southern crops.[3]

By contrast, the House Agriculture Committee again saw an influx of new members. In January 2013 five of nine new Republicans on the committee and eleven of the twelve new Democrats were newly elected (table 8.2). Although a handful of seats on the committee had been vacated due to retirement, defeat, or, in Huelskamp's case, expulsion, most needed to be filled because members had left for seats on more prestigious panels. Democrat Terri Sewell, for example, after only one term in the House, obtained seats on the Financial Services and Intelligence Committees. Even Chellie Pingree (D-ME), who had sponsored her own farm bill in the 112th Congress, left Agriculture for a seat on Appropriations (notably, its subcommittee covering agriculture and rural development), which was still the place to be, despite (or because of) spending cuts. Moreover, Agriculture may not have been a top priority even for those who remained on the committee. For example, twelve returning committee members (seven Republicans and five Democrats) also sat on the Armed Services Committee, a desirable panel among members with military facilities back home.[4] Save for an ever-shrinking number of members from very rural districts, a seat on the House Agriculture Committee, once an icon of clientele representation, no longer held much allure.

The House Agriculture Committee had also become more of a demographic outlier than ever. As Huelskamp and others have observed, by the 1990s, members of the committee already differed from their House colleagues.[5] Regardless of party, they tended to come from more rural districts whose populations were whiter and older than average. This divergence grew wider over the next two decades as the House became increasingly urban and suburban and more diverse. By 2013, there were also stark differences between committee Republicans and Democrats (table 8.3). The Republicans hailed from largely white, non-Hispanic districts in the Midwest, South, and Southwest that had voted for Romney. Their districts may have had farms—which received the bulk of commodity subsidies—but for the most part, their centers of political gravity, and their votes, were in the exurban rings around major cities. As individuals, committee Republicans were all white, mostly male, and uniformly conservative, as measured by a *National Journal* analysis of voting patterns in roll-call votes. They averaged a 74 conservative score in 2013, ranging from 94 for Scott Desjarlais of Tennessee to 49 for Christopher Gibson, whose central New York district was one of only two Republican districts represented on the committee that voted higher than 50 percent for Obama in 2012.

For their part, committee Democrats represented districts that were more urban and less white and whose residents had voted for Obama. Not surprisingly, as individuals, they were less likely to be white (one-third were African

Table 8.2 Members of the House Committee on Agriculture, 113th Congress

Republicans (25)	District	Year First Elected	Democrats (21)	District	Year First Elected
Frank D. Lucas (chair)	OK-3	1994	Collin Peterson (ranking minority)	MN-7	1990
Bob Goodlatte	VA-6	1992	Mike McIntyre	NC-7	1996
Steve King	IA-4	2002	David Scott	GA-13	2002
Randy Neugebauer	TX-19	2003	Jim Costa	CA-16	2004
Mike Rogers	AL-3	2002	Timothy J. Walz	MN-1	2006
K. Michael Conaway	TX-11	2004	Kurt Schrader	OR-5	2008
Glenn Thompson	PA-5	2008	Marcia Fudge	OH-11	2008
Bob Gibbs	OH-7	2010	Jim McGovern	MA-2	1996
Austin Scott	GA-8	2010	Joe Courtney	CT-2	2006
Scott Tipton	CO-3	2010	John Garamendi	CA-3	2008
Rick Crawford	AR-1	2010	Suzan DelBene	WA-1	2012
Martha Roby	AL-2	2010	Gloria Negrete McLeod	CA-35	2012
Scott Desjarlais	TN-4	2010	Filemon Vela	TX-34	2012
Chris Gibson	NY-19	2010	Michelle Lujan Grisham	NM-1	2012
Vicky Hartzler	MO-4	2010	Ann Kuster	NH-2	2012
Reid Ribble	WI-8	2010	Rick Nolan	MN-8	2012
Kristi Noem	SD-at large	2010	Pete Gallego	TX-23	2012
Dan Benishek	MI-1	2010	William Enyart	IL-12	2012
Jeff Denham	CA-10	2010	Juan Vargas	CA-51	2012
Stephen Fincher	TN-8	2010	Cheri Bustos	IL-17	2012
Doug LaMalfa	CA-1	2012	Sean Patrick Maloney	NY-18	2012
Richard Hudson	NC-8	2012			
Rodney Davis	IL-13	2012			
Chris Collins	NY-27	2012			
Ted Yoho	FL-3	2012			

Table 8.3 District Characteristics, House Committee on Agriculture, 113th Congress

	Mean		Median		Mode	
	Republican	Democrat	Republican	Democrat	Republican	Democrat
% Voting for Obama in 2012	40	59	40	54	36	54
% Urban	58	80	56	80	56	81
% White non-Hispanic	77	66	81	70	92	36
Conservative vote rating	77	29	79	35	89	35
Value of subsidies (millions of dollars)	69	33	21	5	8	2
% Receiving SNAP benefits in 2011	13	13	13	12	16	15

Sources: 2010 Census, http://proximityone.com/cd113_2010_ur.htm; American Community Survey, 2009–2011, http://www.fns.usda.gov/Ora/SNAPCharacteristics/default.htm; *National Journal* rankings, Project Vote Smart, https://votesmart.org/interest-group/1868/rating/5953#.V3vM_1ctOcs; Ballotpedia, https://ballotpedia.org/National_Journal_vote_ratings#cite_note-8; USDA, *2012 Census of Agriculture*, table 17, https://www.agcensus.usda.gov/Publications/2012/Online_Resources/Congressional_District_Rankings/cdr_1_017_017.pdf.

American or Latino), more likely to be female (six of twenty), and largely liberal, averaging a 32 *National Journal* conservative rating in 2013.[6] With a few exceptions, their coastal and upper Midwest districts also got far less in commodity subsidies than their Republican counterparts, in part because they had fewer farms overall and because those farms tended to produce crops ineligible for direct subsidies (e.g., fruits and vegetables). The exceptions were the increasingly rare rural Democrats like Collin Peterson of Minnesota and Mike McIntyre of North Carolina, both of whom represented significant farming sectors and scored around 45 on the *National Journal* scale. As such, they were outliers in their own party.

The only area where committee members had some commonality was the mean percentage of households in their districts receiving SNAP benefits—about 13 percent in 2011.[7] It made little difference: Republican Stephen Fincher of Tennessee (whose district's SNAP rate was 20 percent) was vocal in seeking to toughen the rules for SNAP eligibility, while Democrat James McGovern of Massachusetts (15 percent SNAP rate) chaired the Congressional Hunger Center and was on the Agriculture Committee precisely to fight any new restrictions.[8] Fincher, a seventh-generation farmer, would soon be subjected to considerable public criticism after an Environment Working Group analysis of farm subsidies showed that he had received nearly $4 million in direct payments from 1999 to 2012, most of it for cotton.[9]

For Frank Lucas and Collin Peterson, returning as chair and ranking minority member, the committee's worsening outsider status made it harder to guarantee support in the larger chamber and easier for Speaker Boehner to block action on the Farm Bill in 2012, at no cost to his leadership. But, with the extension expiring on October 1, they had to try again. Even as Lucas stated his intent to start committee consideration in February, a still-angry Peterson wrote an open letter to Boehner and majority leader Cantor, demanding that Lucas be given a chance: "I see no reason why the House Agriculture Committee should undertake the fool's errand to craft another long-term farm bill if the Republican Leadership refuses to give any assurances that our bipartisan work will be considered. . . . I believe it is only fair for me to ask for a written commitment that your Leadership team will find floor time during this Congress if the Committee marks-up a new five-year farm bill."[10] There is no record that Boehner or Cantor ever responded.

Once Again into the Breach

Nor had the 2012 elections altered the dilemmas facing the leaders of the committees as they prepared for another go-round. As Lucas observed, "Remember there are three tiers of issues here: committee issues, floor issues and there will be conference issues. I just have to get to conference to sort out some issues."[11] But getting there would be affected by macrolevel budget issues. The automatic cuts threatened by the sequester were delayed only until March 1, 2013, and Congress and the president still needed to reach an agreement on the debt ceiling and the budget, with failure raising the prospect of another shutdown of "nonessential" federal government services. It was, as they say, déjà vu all over again.

House Republicans, riven internally over the just-enacted "fiscal cliff" deal, opted in late January to put off one expected collision by agreeing to "suspend"—in effect, pay no attention to—the debt ceiling through May.[12] Negotiations with the president over the budget continued even as the sequester went into effect on March 1, triggering $85 billion in across-the-board cuts in discretionary spending in fiscal year 2013, including $5 billion in cuts to commodity programs. Entitlement programs, including SNAP, were unaffected. On March 26 Congress finally passed a continuing resolution to fund ongoing programs until October 1, the start of fiscal year 2014. However, the Consolidated and Further Continuing Appropriations Act of 2013 (H.R. 933) left the sequester in place, locking in future years of automatic cuts. At the same time, President Obama proposed a ten-year $1.5 trillion deficit-reduction plan with $38 billion in cuts to USDA funding, nearly all taken from commodity programs. However, beyond laying out spending targets and expressing support for domestic and international nutrition programs, the White House made little effort to shape Farm Bill priorities. Lucas chalked up the administration's hands-off stance to simple politics: "Why worry about the people who are not part of your base, I guess."[13]

In the meantime, as invariably happens, nature had intruded into agricultural policymaking, and the effects of the ongoing drought raised current and projected costs of subsidizing crop insurance.[14] The sudden surge in program costs led the Congressional Budget Office to lower its estimates of savings expected from the 2012 bills: S. 3420 would save $13.1 billion, not the $23.1 billion previously estimated, and H.R. 6083 would cut $26.6 billion, as opposed to the $35.1 billion promised by Lucas. The difference between the two was the Senate's heavier reliance on savings obtained by shifting from direct payments to crop insurance, whereas the House used cuts in SNAP to reach its targets.[15]

For Stabenow and Lucas, these increases in crop insurance costs affected how they proposed to balance regional crop interests while retaining other stakeholders, including nutrition program advocates. Stabenow also had to accommodate Thad Cochran, who was adamant about restoring some form of direct payments in the bill, as favored by rice and peanut growers. To do so and stay within budget targets would require cuts elsewhere, likely in nutrition programs. At least Stabenow knew that Cochran, in contrast to the House Republicans from his state, favored maintaining the longtime "farm programs + food stamps" linkage. Food stamps, Cochran said, "should continue to be included purely from a political perspective. It helps get the farm bill passed." He added, "I come from a state where we have higher-percentage participation. It is part of my representation of the state that I make sure that those interests get represented. I have never had to apologize in Mississippi for supporting it."[16]

Prospects for progress brightened unexpectedly once Congress passed the continuing budget resolution. In the Senate, majority leader Reid expressed a desire to pass the Farm Bill no later than mid-June so that the chamber could move on to immigration reform—a priority among Democrats, not to mention those in agriculture who depended on immigrant labor.[17] At almost the same time, House majority leader Cantor, eager to show some progress after two years of stalemate—and to burnish his own leadership credentials—told Lucas to get a bill to the floor by early summer, even without guarantees of support within the Republican caucus. Everyone wanted bills into a conference committee by August so that Congress could avoid another public failure to reauthorize by the deadline.

Leaders of the committees wasted no time, and over the span of four days in May, they led their panels in coordinated moves to the chamber floors. On May 12 Stabenow and Cochran unveiled S. 954, the Agriculture Reform, Food, and Jobs Act of 2013. As it had a year earlier, the committee's bill promised around $24 billion in savings over ten years, counting the $6 billion already factored into sequestration. However, this time, the distribution of program supports reflected Cochran's leverage, offering direct payments for rice and peanut producers while maintaining insurance programs favored by midwestern grain producers, but with changes in insurance reimbursement rates and schedules to make up-front costs more budget friendly and to keep cuts in SNAP to around $4.1 billion. Stabenow hoped the concessions to southern Republicans would pave the way for floor passage and make it easier to align with the commodity programs expected in the House bill. Stabenow also reluctantly omitted a provision worked out between egg producers and the Humane Society to address the rising number of state-level battles over the treatment of hens. That provision, supported by her state's egg industry, sparked fierce resistance from groups representing midwestern pork and cattle producers, who feared the precedent of Congress dictating housing standards for animals. Their opposition threatened to derail committee action, leaving Stabenow little choice but to abandon the effort.[18]

On May 13—not coincidentally, the day after Stabenow and Cochran unveiled S. 954—Lucas and Peterson proposed H.R. 1947, the Federal Agriculture Reform and Risk Management Act of 2013 (FARRM). The House markup again offered a mix of commodity programs, with greater reliance on countercyclical direct payments than found in the Senate bill, plus a transition program to ease cotton into crop insurance, emergency assistance for livestock producers (close to Lucas's heart), and, again, a dairy management program (also contained in the Senate bill) intended to keep Peterson and other Democrats on board, despite Boehner's known opposition. More important, the chair's mark offered $40 billion in savings over ten years, exceeding the $38 billion sought by the White House. Of that, $6 billion would come from sequestration, $20.5 billion from nutrition programs, and the rest from other titles, conservation in particular.[19] Lucas admitted that this bill was much more dependent on cuts in SNAP than in 2012 but declared, "I sincerely believe this $20 billion won't take a calorie off the plate of anyone who's qualified."[20] Left unstated was whether those cuts would be enough to appease his most conservative colleagues, not to mention the array of outside activist groups readying their lobbying campaigns.

The Senate Agriculture Committee took up consideration of S. 954 on May 14, with members approving the ten-year $955 billion package by a 15–5 vote. S. 954 was similar to its 2012 predecessor, except for the adjustments made in commodity programs to gain the support of southern senators. However, those changes cost Stabenow allies, with midwestern Republicans

Roberts, Thune, and Johanns switching sides because of the rice and peanut programs. Once again, Gillibrand was the lone Democratic holdout because of SNAP cuts. She also proposed but eventually withdrew an amendment to shift funds from crop insurance to SNAP, while several Republican senators failed in their attempts to further narrow food stamp eligibility.[21] McConnell again opposed passage, this time apparently because he could not get Stabenow to include a provision legalizing the cultivation of industrial hemp, an important crop in Kentucky. McConnell, who had previously shown little interest in farm policy, happened to be up for reelection in 2014.[22]

Lucas convened the House Agriculture Committee on May 15, which then debated and passed H.R. 1947 just before midnight that same day. The final vote was a bipartisan 36–10, with vote breakdowns similar to those of the year before: eight no votes from Democrats opposed to cuts in SNAP, and two from Republicans against the dairy program. Along the way, the committee voted on eleven proposed amendments, two with particular relevance to the floor fights to come. As in 2012, Goodlatte's effort to kill dairy management was defeated (26–20), with increased support from fellow Republicans reflecting Boehner's lobbying on the issue. Lucas and five other Republicans stuck with Peterson, making the difference.

The more highly charged conflict was again over nutrition. The committee voted 27–17 against an amendment by McGovern of Massachusetts to eliminate the changes in SNAP eligibility at the core of the $20.5 billion in spending cuts, which, critics charged, would affect nearly 2 million beneficiaries.[23] McGovern argued that tighter limits on household assets and an increase in the LIHEAP threshold would adversely affect the working poor and elderly and that these major program changes should not occur without full committee hearings. "These are not reforms," McGovern charged. "These are just cuts, and they are going to hurt people."[24] Committee Republicans, citing anecdotes about food stamps being dispensed to convicted felons and other able-bodied individuals, argued that their changes would rein in program abuses. Peterson, who had earlier raised the hackles of farm groups by stating that there was far less fraud and waste in SNAP than in some commodity programs, nevertheless sided with Lucas and the Republicans in retaining the new restrictions.[25] Although he would have his hands full keeping liberal Democrats on board when the bill went to the House floor, he also knew that any final version would have to cut less deeply to be acceptable to the Senate or Obama. "At the end of the day this bill is going to be written in conference," Peterson said. "We just need to figure out how to get this to conference. That's the trick."[26]

Of passing note, both committees rejected the Obama administration's only real effort at major policy change: a proposal to move the $1.4 billion foreign food aid program out of the USDA and to loosen requirements that food destined for aid be purchased from US farmers and shipped on US-flagged merchant ships. Critics argued that these longtime rules cost too much, slowed relief efforts, and were little more than blatant subsidies for entrenched domestic interests, the shipping industry chief among them. But farm groups and maritime labor unions, in a rare moment of unity, pushed hard to keep the status quo benefiting their respective self-interests.[27]

Reid Keeps the Mojo Going

There's an old saying about legislatures: "Everything has been said, but not everyone has said it."[28] That adage certainly applied to the Senate's consideration of S. 954, which, save for accommodations made to southern crops, differed little from its 2012 predecessor. Still, this being the Senate, floor consideration once again offered all 100 senators an opportunity to propose amendments and to talk about them.

But Harry Reid was in no mood for a lengthy debate over a bill that had passed easily the year before and that he needed to get off his to-do list as soon as possible. He quickly set a deadline by which Stabenow and Cochran had to get sixty votes for cloture, and he made it clear that it was now or never. They had two weeks, over the Memorial Day recess, to wade through more than 240 possible amendments, most of which were repeats from the year before.

Floor action on S. 954 began on May 20, with Reid again allowing votes on several litmus-test amendments while Stabenow and Cochran gathered their sixty votes. Most were replays of battles fought in 2012. Of note were three amendments on food stamps, two filed by Agriculture Committee members. The first, by Kirsten Gillibrand, would replace the bill's cuts in SNAP with equal cuts in crop insurance. She lost 70–26, with veteran Democrats Dianne Feinstein of California and Barbara Mikulski of Maryland siding publicly with Stabenow against their first-term colleague.[29] A second, by Pat Roberts, would cut SNAP by an *additional* $31 billion—more than called for in the

House bill. Observers believed that Roberts, no longer constrained as the committee's ranking minority member, was looking to stave off a primary challenge from his right as he neared reelection in 2014. He also lost, 58–40, with three Republicans—including Cochran—joining all the Democrats in opposition.[30] Finally, an amendment by Republican James Inhofe of Oklahoma to turn SNAP into a state block grant program failed 60–36, with nine Republicans—including Chambliss, Cochran, and Roberts—joining Democrats in opposing what most saw as a poison pill guaranteed to prompt a White House veto.[31]

Critics of the sugar program once again got their chance and lost by nearly the same margin as before. The outcome was marked by one classic coalition of strange bedfellows: senators from sugar-producing states—including Tea Party Republican Marco Rubio of Florida (sugarcane) and liberal Democrat Debbie Stabenow of Michigan (sugar beets)—defeating an equally odd coalition of senators from sugar-consuming states—including liberal Democrat Richard Durbin of Illinois, home of Brach's chocolates, Wrigley's gum, and Tootsie Rolls, and Tea Party Republican Patrick Toomey of Pennsylvania, home of Hershey's chocolates.[32] Bernie Sanders of Vermont once again proposed that states be allowed to mandate labeling on food products with genetically modified ingredients. In a surprise switch from 2012, Reid and Chuck Schumer of New York, widely seen as Reid's eventual successor as Democratic leader, supported Sanders in what proved to be a 27–71 losing effort. All but one aye vote came from Democrats from heavily urban states, while Republicans, echoing industry concerns about the proverbial "checkerboard" of state labeling laws, were joined in opposition by Stabenow and most farm-state Democrats.[33]

The rare successful floor amendment of any note, by Democrat Durbin of Illinois and Republican Tom Coburn of Oklahoma, reduced crop insurance premium subsidies by 15 percent for farmers earning $750,000 or more. The committee bill sought to negate the need for income tests by tying insurance subsidies to compliance with conservation measures and by tightening the definition of being "actively engaged" in farming, but as in 2012, senators wanted to be tough on large producers. The amendment passed 59–33, despite opposition by Stabenow and Cochran.[34]

On June 4, after Congress's return from the Memorial Day recess, Stabenow signaled that she and Cochran thought they had their sixty votes, even though more than 200 proposed amendments had been left unresolved. Reid filed for cloture, allotting only thirty minutes for debate. Still uncertain was to what extent midwestern Republicans would oppose cloture because of Cochran's apparent refusal to let them try to eliminate market-distorting (in their opinion) countercyclical payments for southern crops.[35] In the end, most senators decided that moving a Farm Bill was better than not doing so, and two days later the Senate imposed cloture by a 75–22 vote, with midwestern Republicans making up the bulk of the no votes.[36]

The Senate passed S. 954 on June 10 by a 66–27 bipartisan majority, a higher margin of victory than in 2012.[37] Of note, eighteen Republicans, including several from southern states, joined Democrats in the majority. All but two no votes were by Republicans, including Roberts and McConnell; as in 2012, Democrats Reed and Whitehouse of Rhode Island voted no to protest cuts in SNAP.[38] Stabenow, recalling the frustrations of the previous year, spent little time exulting in the Senate's passage of the bill: "Hopefully the House this time will complete this work and we'll have an opportunity to go to conference."[39] Speaker Boehner concurred: "I'm hopeful that we can pass a Farm Bill and get to conference with the Senate and resolve this issue for America's farmers and ranchers."[40]

Farm Bill—or Food Stamp Bill?

Hopes that the Senate's vote on S. 954 would spur the House soon sagged as Frank Lucas and House leaders judged the scope of Republican floor support for H.R. 1947. Whatever other obstacles lay before them—notably, Boehner's opposition to Peterson's dairy program—the deepest fault line remained food stamps. On one side were many, if not most, members of the Republican caucus, typified by Tim Huelskamp, for whom the Farm Bill had lost its traditional meaning. "Unless there are some significant changes and some significant effort, I don't think this'll get across the floor," Huelskamp said soon after the Senate's action. "People in Kansas get it—just because it's the farm bill doesn't mean you have to vote for it. It's not just the farm bill, it's also the food stamp bill.[41] Members like Huelskamp were backed—critics argued, intimidated—by a phalanx of well-funded advocacy groups instrumental in the Tea Party insurgency and determined to hold conservatives to their campaign promises. To cite one example, Heritage Action, the political action wing of the Heritage Foundation, aimed a district-level radio campaign against the Republicans on the House Agriculture Committee, casting H.R. 1947 as little more than a food stamp bill. "Only 20 percent of the funds would go to support farmers," the ad declared. "The rest would go to bankroll President Obama's food stamp agenda."[42]

On the other side were nutrition advocates and their allies in the Democratic Party. Collin Peterson, the lead Democratic negotiator, knew that SNAP's defenders had no incentive to compromise because spending on the entitlement program would continue even if the 2008 law expired. His job was to help Lucas keep SNAP cuts to $20.5 billion and hope that a conference committee could get closer to the $4 billion in the Senate bill, which might be acceptable to President Obama. Peterson did not expect more than 150 Republicans to support the bill as configured, and he knew that Boehner would need at least seventy Democrats to offset Republican defections.[43] But with fewer rural Democrats than ever, his margin of error was slim. As it was, 134 out of 201 House Democrats had already signed a resolution opposing *any* cuts in SNAP.[44] Said McGovern of Massachusetts, "I hope no one votes for it. If we don't stand for the poor and vulnerable, what do we stand for?"[45] For her part, minority leader Nancy Pelosi dared Boehner to pass the bill with only Republicans: "I know how hard this is to weave together," she said, reflecting her experience as Speaker in 2008. "You don't start by having $20 billion in SNAP cuts unless you decide you are going to have 218 votes to pass a bill."[46] Pelosi, though sympathetic to Peterson's task, would join her party's majority in opposition unless the additional cuts in SNAP were taken out.

Lucas and Peterson spent early June trying to round up the 218 votes they needed, reminding skeptical colleagues that SNAP spending would continue without reauthorization, while many of their favored farm and food programs—including some of interest to urban Democrats—would not.[47] Peterson by now hoped to keep around forty Democrats in the fold, largely those who supported his dairy program and provisions for specialty crops and healthy eating. With those Democratic votes, Lucas felt mildly optimistic about the prospects for passage, even given regional splits among farm bloc Republicans over commodity programs and expected defections by conservatives opposed to the bill's overall spending. Finally, Boehner announced his intent let H.R. 1947 go to the floor even though there was no apparent consensus within his own caucus. The price of violating the Hastert Rule would be uncertainty about the outcome: the leadership, through the Rules Committee, would allow the consideration of nearly 100 floor amendments to give various factions a chance to be heard. Boehner also declared that he would vote for whatever version the majority decided on.

On June 19 the House took up the first of the amendments, tellingly, one by McGovern to reverse the $20.5 billion in SNAP cuts. It failed by a vote of 234–188, almost entirely along party lines, with a handful of rural Democrats, including Peterson, joining the majority.[48] "We're not going after big agribusiness. We're not going after crop insurance. What we're doing is we're going after poor people," McGovern charged, to which Pete Sessions (R-TX) responded, "We're trying to help prioritize and save this system."[49] Seeking to help Peterson keep enough Democrats in the fold, House leaders persuaded Michael Conaway (R-TX) to allow a voice vote on a largely symbolic amendment to impose a 10 percent cut in SNAP spending should the Farm Bill not be reauthorized.[50] It failed, but soon thereafter the House approved by voice vote a so-called en bloc amendment that included a lifetime ban on benefits to certain convicted felons and permission for states to require drug testing for recipients. To many observers, these were little more than "bumper sticker" provisions, more symbolic than practical in effect, but their adoption only heightened Democrats' animosity toward the larger bill.[51]

Lucas, aided by a coalition of southern Republicans and sixty Democrats, narrowly (217–208) fought off an amendment offered by Democrat Ron Kind of Wisconsin and backed by midwestern corn and soy producers to cap countercyclical payments and impose an income means test on any direct subsidies.[52] Again with the help of sixty Democrats, Lucas also managed to defeat (230–194) an effort by Republican Jeff Fortenberry of Nebraska to impose a payment cap for any one farm and to clarify the meaning of "actively engaged" in agriculture.[53] Other amendments aimed at commodity programs were withdrawn as supporters decided to save their arguments for conference. For Lucas and Peterson, so far, so good.

The expected battle over the dairy program took on new weight when Boehner, in a rare move, distributed a "Dear Colleague" letter outlining his opposition to Peterson's plan and supporting a Goodlatte amendment to replace supply management with an insurance program favored by larger producers. Boehner's lobbying, and claims that supply management would mean higher consumer prices, persuaded most Republicans and half of the

Democrats to support the amendment (291–135).[54] Even though Peterson had the support of Democrats from the Northeast and upper Midwest, he could not withstand Boehner's pressure—accentuated by his vote, a rare instance of a sitting Speaker voting on a floor amendment. The final battle over commodities was yet another effort to eliminate the longtime sugar program. It lost 221–206, with both parties split on the issue.[55]

The sugar vote set the stage for two final floor amendments, both over SNAP, which House leaders had scheduled last in the hope that the opportunity to vote on major program changes might appease conservatives. The first amendment, by Huelskamp, would impose new federal work requirements on recipients and deepen program cuts to $31 billion. It failed 250–175, with 57 Republicans—including Lucas and others on the Agriculture Committee—joining all the Democrats present in opposition.[56] Even though Lucas had tried to persuade his colleagues not to support an amendment he knew would sink the entire bill, majority leader Cantor, majority whip Kevin McCarthy of California, and other members of the Republican leadership team were notable in their support of it.

Immediately after came the last floor amendment, a far-reaching and complicated package sponsored by Steve Southerland (R-FL) to give states the option to impose stricter work requirements on SNAP recipients—and to let the states keep some of the savings when benefits were terminated. Supporters argued that these requirements were essential to encourage recipients to work, while opponents saw the new rules as little more than additional and punitive hurdles for people already struggling to find jobs in a period of high unemployment. Cantor, widely known as the mover behind the Southerland amendment, went to the floor and spoke on its behalf, much to the dismay of Agriculture Committee leaders, who were anxious about losing more Democrats. The amendment was approved by voice vote, at which time Southerland pointedly demanded a roll call to make members' votes public.[57] The outcome was a near party-line vote, 227–198, with Jim Cooper of Tennessee the lone Democrat voting yes. Announcement of the vote tally elicited audible gasps in the House chamber.[58]

If Republican leaders thought the Southerland amendment would appease conservatives enough to vote for the final bill, they were quickly proved wrong. Following a final plea by Lucas that his colleagues overcome their differences to move the bill to conference—which earned him a standing ovation from members of both parties—the House immediately voted 234–195 against H.R. 1947, with 62 Republicans joining all but 24 Democrats in opposition. To the Democrats, who were already furious about what they considered a punitive campaign against food stamp recipients, the Southerland amendment was the last straw; at least two dozen who had pledged their support to Peterson changed their minds the moment the amendment passed. For their part, after supporting the amendment, 58 Republicans, including Huelskamp, voted against final passage because the bill did not go far enough to cut spending.[59] Boehner, Cantor, and McCarthy all voted yes, but it was not enough.

Members watched, stunned, as the votes registered on the chamber's electronic tally boards and a bill that most expected to pass went down to defeat. For Lucas, Peterson, and the House leadership, it was a disaster. Democrats pointed fingers at Cantor and McCarthy for not locking in party support for the bill, noting that seven Republican no votes had come from other committee chairs, a telling and in some ways unprecedented breach in party discipline. "I just can't get over the fact that 58 Republicans voted for an amendment that would sink the bill," said Pelosi, reminding everyone that President Obama had warned he would veto the bill if it contained the threatened SNAP cuts. "It's a stunning thing. Why would you give people an amendment that's going to kill your bill, and then go blame it on somebody else?"[60] Republicans in turn blamed Pelosi and Peterson for not delivering the promised Democrats, a charge to which minority whip Steny Hoyer of Maryland retorted, "We take no blame for the farm bill. None. Zero." Peterson singled out Cantor for allowing the Southerland amendment to go to the floor despite his warnings that it would be the proverbial last straw for already angry Democrats. "When I was chairman, I had to come up with the votes," he said, reminding observers that he had pushed his original version of the 2008 bill through the House with only nineteen Republican votes. A dejected Lucas simply said, "This turned out to be an even heavier lift than I thought."[61]

A Divorce of Convenience

Farm groups expressed disappointment at another failure, but their sentiments were shared by relatively few others. The editorial board of the *Wash-*

ington Post opined, "Pardon us for not rending our garments at the downfall of a measure that would have lavished tens of billions of dollars in subsidies on one of the most prosperous sectors of the U.S. economy while cutting $20 billion over 10 years from a major program for the poor."[62] The conservative Heritage Action applauded the votes of House Republicans like Marlin Stutzman of Indiana, who said, "While it might have been called a 'farm bill,' the American people understand that it was anything but. This trillion-dollar spending bill is too big and would have passed welfare policy on the backs of farmers." They were joined by critics on the Left, such as the Environmental Working Group, which called H.R. 1947 "a bloated farm bill that increases subsidies for the largest and most successful farm businesses, while needlessly cutting programs designed to help feed the hungry and protect the environment."[63] The Farm Bill had failed, and much of America didn't care.

Even so, House leaders immediately looked for ways to salvage the situation, if only to deflect criticisms that they had lost control. Trial balloons for another one-year extension met with scorn, and Harry Reid was particularly blunt: "I want everyone within the sound of my voice—as well as my colleagues on the other side of the Capitol—to know that the Senate will not pass another temporary farm bill extension."[64] Another option, to renew program spending through the Appropriations Committee, met with threats of points of order from Lucas, who was not inclined to let House leaders use the appropriations process to legislate matters more properly left to his panel.

House leaders thereafter abandoned any pretense that the "farm programs + food stamps" linkage still served its traditional purpose. Cantor, over strenuous objections by Lucas and Peterson, moved nutrition programs into a separate bill so that conservatives could vote for a "stand-alone" Farm Bill without having to support food stamps.[65] His move, echoing tactics by Senate Agriculture Committee chair Jesse Helms decades earlier, was applauded by conservative advocacy groups such as Heritage Action and the Club for Growth as a way to isolate and cut back an out-of-control SNAP, and by commodity groups that were worried about the fate of their own programs as fighting over SNAP continued to block reauthorization. But that tactic was opposed by a coalition of 532 agricultural groups, including the normally conservative American Farm Bureau Federation, and it was seen as risky by Farm Bill veterans, given the lack of strong support for commodity programs among most House members. As if to accentuate the riskiness of Cantor's strategy, President Obama promised to veto any bill that did not contain nutrition programs.[66]

On July 11 the House passed H.R. 2642, a "farm only" version of the Federal Agriculture Reform and Risk Management Act of 2013. The final vote, 216–208, was almost entirely along party lines, with all 196 Democrats voting against it. Also voting no were 12 Republicans, including Huelskamp.[67] To its critics, the stand-alone version embodied the worst impulses of commodity producers, with budget savings coming almost entirely from conservation programs. Of note, H.R. 2642, which authorized $196 billion in spending over ten years, contained provisions repealing the 1938 and 1949 laws on which commodity program authority had long rested, apparently because many commodity producers, southern sugar, rice, and cotton producers in particular, had grown tired of the uncertainty that came with each five-year reauthorization.[68] Peterson warned his colleagues that repealing the sunset provisions in agriculture's "permanent law" would eventually backfire:

> If you want to ensure Congress never considers another farm bill and the farm programs as written in this bill remain forever, then vote for this bill. In every farm bill, there are things some people like and things some people don't. The beauty of the '38 and '49 permanent laws is that it forces both groups to work together on a new farm bill, because no one really wants to go back to the old commodity programs.[69]

In one of those "politics makes strange bedfellows" moments, conservative and liberal critics of the long-standing commodity support system echoed Peterson's objections. They would fight to retain the opportunity to revisit agricultural policy every few years.[70]

Republican leaders expressed satisfaction at passing the first "pure" Farm Bill in forty years. Lucas, the good soldier, argued that those Democrats concerned about nutrition programs should be pleased with the separation. "The way food stamps is structured, even if the authorization expires, it's an appropriated program anyway. They can just keep it going," he said. But McGovern, evoking the bitter experience with welfare reform in 1996, did not buy that argument and responded that the Appropriations Committee "could take a meat ax to the program. I don't believe they will allow benefits to continue at the current level."[71]

Over in the Senate, Stabenow moved quickly to seek a conference, regardless of any House action on nutrition programs. On July 18 she obtained unanimous consent for the Senate to take up H.R. 2642, as amended by the text of S. 954, and insisted on a conference.[72] House leaders made no effort to respond. Instead, they were struggling over what to do about nutrition programs, with Cantor and conservatives seeking to put together a package making major—in critics' eyes, draconian—changes in SNAP eligibility and program implementation. Lucas was unable to get Boehner to act on the Senate request for a conference until Cantor had been given a chance to move on the SNAP bill, which would not come until after the August recess. Lucas and Peterson both worried that the animosities stirred up by the bitter fight over SNAP might cause collateral damage to their efforts on the Farm Bill.[73] In the meantime, Secretary Vilsack made it clear that President Obama would veto another one-year extension.

On September 19 Cantor—bypassing the committee markup process and allowing no floor amendments—pushed through the Nutrition Reform and Work Opportunity Act of 2013 (H.R. 3102) by a 217–210 party-line vote, with Republican leaders forced to pull in members who had planned to vote no to offset fifteen defections. H.R. 3102 contained many of the "reforms" sought by conservatives in the fight over H.R. 2642, leading to $39 billion in SNAP savings over ten years—ten times that in the Senate bill.[74] Of note, H.R. 3102 authorized SNAP for three years, compared with the five-year authorization for commodity programs in H.R. 2642. The Senate refused to adopt either House bill as a substitute for S. 954 and awaited a formal House request for a conference—and waited.

Finally, on September 28 the House, by a 226–191 party-line vote, agreed to S. 954 as amended by H.R. 2642 *and* with a new Title IV incorporating the text of H.R. 3102.[75] The "farm programs + food stamps" linkage was renewed, in no small part because House leaders knew that that no bill would get past the Senate—or Obama—without it. Whatever the later political ramifications, Cantor's tactical separation of the two had paid off. The bill was headed to conference.

Three days later, on October 1, authorization for the 2008 law expired, again. For most Washington observers, however, that wasn't the big news. Also on October 1, the whole government shut down.

9

In Conference

It shouldn't be this hard. But we live in a time when things are hard so we just have to work a little harder.
—Frank Lucas

Remember the debt ceiling? Back in January, House Republicans had opted to "suspend" the nation's borrowing limit through mid-May while they wrangled with Senate Democrats and the administration over fiscal year 2013 spending, reaching a deal on current funding in March. However, stalemate continued, and the nation technically hit the debt limit on May 19, largely due to the Republicans' determination to repeal the Affordable Care Act (ACA). The Treasury Department announced that the nation would not start to default on its obligations until sometime in October, just after the start of fiscal year 2014. Congress and the president also continued to wrangle over the fiscal year 2014 budget, with House (and a few Senate) Republicans determined to use the debt ceiling as leverage in their efforts to defund Obamacare and impose deeper cuts in federal spending.

In late September the House passed a continuing resolution to extend fiscal year 2014 funding through November 2013, which the Senate also passed after stripping out provisions defunding ACA implementation. Senate Republicans chose not to block that action to avoid being blamed for a possible government shutdown. But House conservatives were adamant in their demands, overruling

Boehner's own efforts to avoid a shutdown, and the two chambers failed to reach an agreement by midnight on September 30.[1] On October 1, with no authorized spending for fiscal year 2014, the federal government suspended "nonessential" services, ranging from routine visa processing to, more visibly, the closure of national parks and war memorials, and it furloughed nearly 1 million federal workers.[2] Ironically, the first open enrollment period for the ACA started as scheduled; implementation of its health insurance provisions—a mandated program with separate funding—was unaffected.

The shutdown coincided with the expiration of the 2008 farm bill, a convergence that had immediate impacts on an array of USDA programs, including a cessation of technical reports offered by the Economic Research Service, closure of lands managed by the US Forest Service, delays in processing disaster payments to South Dakota ranchers who had lost livestock in a freak early snowstorm, and a cutoff in benefits for the Women, Infants, and Children (WIC) nutrition program.[3] At the same time, the USDA again faced a "dairy cliff," resulting from mandates in the permanent law that it begin buying bulk milk at 1949 prices as soon as January 2014—assuming it had any money to do so.

Opinion polls showed widespread public disapproval of the shutdown, even among Republican voters, putting pressure on party leaders to make a deal. The shutdown lasted just over two weeks, ending just as the nation was about to start defaulting on its debts. Congress agreed to another short-term deal that extended current spending to January 15, 2014, and the debt limit until February 7. Budget sequestration continued as before, ensuring that there would be more automatic budget cuts. House approval garnered the votes of all 198 Democrats and only 87 Republicans, including Boehner and Cantor. The end of the impasse, even if temporary, was seen as an embarrassment for conservatives, as the deal included almost none of their demands.[4]

Through it all, those inside and outside Congress who were eager to get back to the Farm Bill could only wait, as farm and commodity groups grew increasingly concerned about the long-term impacts of another failure to act. Trial balloons related to including a Farm Bill extension in a budget deal were floated and quickly shot down. Lucas and Stabenow in particular wanted nothing to do with a possible repeat of McConnell's dealings the previous December.[5] For all, then, the end of the shutdown came as a relief: the House and Senate could finally get into conference.

Scenes at an Arranged Marriage

Article I, section 7, of the US Constitution stipulates that the House and Senate must concur on a single version of a bill before it goes to the president. It does not specify *how* they are to concur. Sometimes one chamber simply accepts the other's version; at other times there is a short back-and-forth amendment period until agreement is reached. But, as is often the case with particularly complex and critical pieces of legislation, sometimes the chambers stick to their respective guns and agree to form a conference committee.

Rules for conferences seem straightforward: each chamber appoints an unspecified number of conferees and charges them with negotiating a compromise between the two versions, sometimes with precise instructions on what changes the chamber is willing to accept. Each chamber also has rules on the extent to which its conferees can agree to changes or accept new language. For example, House conferees cannot agree to nongermane amendments. A majority of conferees in each chamber must agree to any changes and on the final version. Committee sessions are typically closed, and most negotiations tend to occur among a select group of senior-level principals, or managers, who consult with their respective sets of colleagues throughout the process.[6] Whatever version the conference agrees on goes to the two chambers for a simple up or down vote, with no opportunity for amendment.

Farm bills almost always go to conference. In fact, since 1933, the *only* such bill to avoid one was the game-changing 1973 law shaped by Earl Butz, during which the chambers had a short series of exchanges before concurring on the final version. As a result, conferences on farm bills have developed their own routines: each chamber's conferees tend to come from its Agriculture Committee and are led by the committee chair, with the more senior of the two chairs managing the conference overall. But party leaders in each chamber are free to name additional conferees, whose identities reveal a lot about the controversies of the moment.

There was little question that Harry Reid would follow tradition, and all twelve Senate conferees came from its Agriculture Committee (see table 9.1). The real question was how John Boehner would balance the traditional views represented on the House Agriculture Committee with those of conservatives who wanted major changes and deeper spending cuts, particularly in

Table 9.1 Conferees on H.R. 2468

House of Representatives		Senate	
Republicans (17)	Democrats (12)	Republicans (5)	Democrats (7)
Frank Lucas (OK)*	Collin Peterson (MN)*	Thad Cochran (MS)*	Debbie Stabenow (MI)*
Steve King (IA)*	Mike McIntyre (NC)*	Saxby Chambliss (GA)*	Patrick Leahy (VT)*
Randy Neugehauer (TX)*	Jim Costa (CA)*	Patrick Roberts (KS)*	Tom Harkin (IA)*
Mike Rogers (AL)*	Timothy Walz (MN)*	John Boozeman (AR)*	Max Baucus (MT)*
Michael Conaway (TX)	Kurt Schrader (OR)*	John Hoeven (ND)*	Sherrod Brown (OH)*
Glenn Thompson (PA)*	Jim McGovern (MA)*		Amy Klobuchar (MN)*
Austin Scott (GA)*	Suzan DelBene (WA)*		Michael Bennet (CO)*
Rick Crawford (AR)	Gloria Negrete McLeod (CA)*		
Martha Roby (AL)*	Filemon Vela (TX)*		
Kristi Noem (SD)*	Eliot Engel (NY)		
Jeff Denham (CA)*	Sander Levin (MI)		
Rodney Davis (IL)*	Marcia Fudge (OH)*		
Edward Royce (CA)			
Tom Marino (PA)			
Dave Camp (MI)			
Sam Johnson (TX)			
Steven Southerland (FL)			

Note: Members are listed in their order of appointment by the respective lead conferees.

* Agriculture Committee member.

SNAP. In the end, ten of seventeen Republicans and ten of twelve Democrats named as House conferees had seats on the Agriculture Committee. Notable in his absence was Bob Goodlatte of Virginia. Observers believed that Boehner had omitted Goodlatte to strengthen Lucas's position in conference and because Goodlatte had voted against H.R. 1947 in June. Boehner would keep Lucas on a short leash when it came to any deals, particularly on the dairy program. Committee Democrats included Jim McGovern of Massachusetts and Marcia Fudge of Ohio, chair of the Congressional Black Caucus, both ardent defenders of SNAP. Their inclusion by minority leader Nancy Pelosi would limit Collin Peterson's range of action when making deals on nutrition programs. More telling was the addition of five noncommittee Republicans, including Steven Southerland of Florida, named at the behest of majority leader Eric Cantor to represent conservative views on SNAP reform. An interesting wrinkle was the addition of House Foreign Affairs Committee chairman Edward Royce (R-CA) and ranking member Eliot Engel (D-NY), both of whom favored Obama administration proposals to reform international food aid programs.[7]

The twenty-nine House conferees named in 2013, versus only fourteen appointed in 2008, was seen as evidence of Boehner's need to accommodate factions within his own party.[8] As if to underscore that point, in passing the resolution to name conferees, Republicans barely (204–195) defeated a motion by Peterson to instruct House conferees to use a five-year timeline in reauthorizing both farm programs and food stamps, not the three-year timeline for SNAP put in place by H.R. 3105. Peterson later expressed his hope that Boehner and Cantor would let Lucas and the House conferees do their jobs without interference.[9]

Lines of Departure

Formal conference deliberations did not start until October 30, largely because Reid gave senators a week off to recover from the budget and debt ceiling negotiations. Lucas, based on his seniority, would chair the conference.[10] In the committee's only open meeting, members took the opportunity to make their opening remarks—to "vent," in the words of one conferee. The four Agriculture Committee leaders expressed their hope that the conference could maintain control of the process and finish before late November—the deadline to get a bill through Congress by the end of 2013 to avoid yet another "dairy cliff" and, just as important, to avoid getting entangled in the ongoing struggles over the budget. "The Budget Committee will not be writing the farm bill," declared Debbie Stabenow. "We will write, we will edit, we will offer responsible cuts."[11]

The definition of "responsible" was, as always, open to interpretation. But, unlike in 2008, when conferees had built what proved to be a veto-proof coalition simply by providing new funding for organics and specialty crops, keeping direct cash payments for a range of commodities, and expanding SNAP eligibility, in 2013 they had to figure out how to make enough cuts to satisfy fiscal conservatives while providing sufficient program supports to keep everyone else on board. Based on signals from the White House, they also knew that the president was likely to accept whatever Congress was able to approve, so long as it did not include the most controversial changes in SNAP.

Conferees would wrangle over a range of key differences between the respective bills, as table 9.2 shows.

SNAP. Residing in the background during negotiations was a major change in SNAP that had nothing to do with the Farm Bill: an end, on November 1, to a temporary increase in benefits provided by the 2009 economic stimulus package. The effect was an immediate $5 billion cut in SNAP spending, with another $6 billion decrease in projected future costs. Though expected, the roughly 6 percent decrease in supplemental benefits for recipients was a matter of consternation for nutrition program advocates, and it stiffened their resolve against additional cuts. The White House echoed their concerns. Stabenow publicly agreed, arguing that the projected $11 billion reduction should be added to the equation measuring the food stamp cuts in both bills. "That $11 billion plus the $4 billion in cuts in the Senate bill means that accepting the Senate nutrition title would result in a total of $15 billion in cuts in nutrition," she said.[12] Stabenow also noted that the Congressional Budget Office (CBO) projected lower SNAP costs as the economy improved and more recipients got jobs. Lucas, mindful of the gap in the House and Senate bills, thought that any resolution on SNAP might require negotiations among Obama, Boehner, and Reid as part of yet another last-minute deal on the budget—a prospect that nutrition program advocates feared.[13]

Commodity Supports. Corn prices, meanwhile, were experiencing their

Table 9.2 Major Differences between House and Senate Farm Bills

Provision	H.R. 2642 + H.R. 3102	S. 954
Projected 10-year budget with savings over 2008 Farm Bill baseline costs*	$921 billion; $58 billion in cuts ($39 billion from SNAP)	$955 billion; $24 billion in cuts ($4 billion from SNAP)
Commodity titles	Repeal of the 1938 and 1949 titles; would make 2013 titles permanent law	Retention of 1938 and 1949 titles as permanent law
Subsidy restrictions	House resolution agreeing to Senate floor amendment but with a $950,000 3-year adjusted gross income; looser rules on definition of "actively engaged in farming"	Floor amendment to reduce subsidies for farmers who make more than $750,000 a year; conservation compliance provisions; tighter rules on definition of "actively engaged in farming"
Commodity supports	Focus on covering production costs; price loss coverage (PLC)—countercyclical payments on costs over target prices; supplemental coverage option (SCO)—"shallow-loss" insurance program Most critical supporters: SCO—Lucas, Peterson, southern Republicans; PLC—Midwest Republicans	Focus on protecting revenues; agricultural risk coverage (ARC)—"shallow-loss" payments on insurance; adverse market payment (AMP) plan with targeted payments Most critical supporters: ARC—Midwest Republicans, soy and corn; AMP—southern Republicans, peanuts and rice
Dairy "management" program	Included in committee bill, deleted on House floor Most critical supporters: Northeast and Midwest Democrats	Included in committee bill and final Senate version Most critical supporters: Northeast and Midwest Democrats

(continued)

Provision	H.R. 2642 + H.R. 3102	S. 954
SNAP	Would reauthorize program for 3 years with $39 billion in cuts, significant changes in program eligibility and implementation; would disallow state waivers on work requirements Most critical supporters: House Republicans	Would reauthorize program for 5 years with $4 billion in cuts, modest changes in program rules Most critical supporters: Senate Democrats
Other points of conflict	Repeal of country-of-origin labeling (COOL) rules for meat	No similar provisions
	Would make it illegal for a state to ban food from another state produced under conditions the receiving state found unacceptable	
	Repeal of funding for a USDA catfish inspection station	

Sources: Based on various sources; for a detailed side-by-side comparison, see Ralph M. Chite et al., *The 2013 Farm Bill: A Comparison of the Senate-Passed (S. 954) and House-Passed (H.R. 2642, H.R. 3102) Bills with Current Law* (Congressional Research Service, 2013).

*Includes $6.4 billion in automatic cuts from sequestration over the period, all from agriculture programs (SNAP was exempt from sequestration).

greatest one-year drop in more than fifty years, owing partly to a softening in the biofuels market that had fed its boom. This led to charges that already well-off corn producers (or their landlords) would unduly benefit from a commodity title reflecting the Senate bill's reliance on "shallow-loss" revenue coverage. To Lucas—who was always ready to remind his colleagues that farm programs were meant to protect from "real" losses, like those that had hit High Plains cattle producers earlier that fall—the prospect of compensating growers for "shallow" losses created by market fluctuations only reinforced his determination to retain some type of countercyclical payments and to raise income thresholds for crop insurance subsidies.[14] "There were two different camps on how we should proceed and they were absolute," Lucas later observed. "Corn is not only the biggest volume crop raised in this country, it is the crop that drives all other grain prices. . . . That means there's a sense of urgency among people who might otherwise have said, 'I don't need all of the various features, I may not even need a farm bill. I'm taken care of.' But now suddenly as the price of corn comes down?"[15] But Lucas knew the conferees could not satisfy everyone simply by stapling the House and Senate approaches together. The budget mattered, and the specifics of any deals on commodity supports, such as whether to base subsidies on acres owned (or "base acres") versus acres actually planted, hinged on how the CBO scored their respective projected costs.

Don't Forget Milk. Both bills replaced a range of dairy supports contained in the 2008 law with a direct payment program triggered when national average production costs exceeded national average milk prices. Changes in the milk program were so numerous and so arcane that they filled more than ten pages of small print in the Congressional Research Service's side-by-side analysis of the two bills.[16] The key difference was the "management" program contained in the Senate bill but deleted from H.R. 2642 by the Goodlatte amendment, which would require participating producers to reduce output when margins between costs and target prices were low. Observers did not see Speaker Boehner budging on this issue; nor did they expect Peterson, backed by northeastern and midwestern Senate Democrats, to give much ground.

Other Stuff that Might Kill the Bill. The bills also differed on matters that reflected their divergent constituencies. One critical point of contention was "country of origin" labeling (COOL) for meat, a provision in the 2002 Farm Bill promoted by organic and artisanal beef producers to assure American consumers of the quality and safety of their meat. The COOL provision had always been opposed by large-scale midwestern beef producers that sourced some of their livestock from countries such as Mexico and Canada. Over time, it had become a matter of international tension as Mexico and other US trade partners complained that it created unfair perceptions about their products. An adverse World Trade Organization (WTO) ruling in 2012 led the USDA to promulgate new COOL rules in May 2013 that it hoped would pass WTO muster and give US consumers the information they wanted. Lucas and other House members, at the behest of the National Cattlemen's Beef Association and other producers, sought to require the USDA to produce a report on COOL's impact on domestic meat and poultry prices. The Senate bill contained no comparable provision, reflecting COOL's popularity with consumers, as well as with senators from the upper Plains states who supported their independent ranchers.

The House bill also contained an amendment by Steve King (R-IA) to prohibit states from imposing food safety and animal welfare regulations on food produced in other states. The amendment, proposed at the behest of midwestern egg, meat, and poultry producers, came in response to California's efforts to impose uniform treatment standards for animals, such as minimum cage sizes for egg-laying hens, on producers located in other states. The King amendment was opposed by consumer and animal welfare groups and by most Senate Democrats. Another House provision repealed part of the 2008 law that shifted the inspection of imported farm-raised catfish from the Food and Drug Administration (FDA) to the USDA. Its supporters saw the new facility as redundant and a blatant effort to create barriers to the importation of catfish from Southeast Asia. Its opponents included Thad Cochran and other senators in the southern catfish-farming region.

Finally, the House version repealed the 1938 and 1949 commodity titles, replacing them with a new permanent law that contained no sunset provisions. That change had been pushed by commodity groups tired of uncertainty about farm subsidies, but it was opposed by most general farm groups, including the usually contrary American Farm Bureau Federation and the National Farmers Union, both of which worried that a new permanent law would simply lock in the advantages of those that already enjoyed favored status.[17]

Whatever It Takes—the Conference Round

Talks among the principal conferees—Lucas, Stabenow, Peterson, and Cochran—continued nearly daily through November, even as the House adjourned for two weeks and other business swirled about them. Sometimes they met with staff present; sometimes it was just the four of them, alone, when talks got especially heated. Although these four had been granted discretion in their negotiations, they were mindful of and constrained by the need to represent their respective chambers and constituencies. Lucas and Peterson had the narrowest paths to navigate, often to the frustration of Stabenow and Cochran. Given the private nature of their discussions, their every public word—or the lack thereof—was followed closely by outside stakeholders keen to know where all this was heading.

Not that the outsiders were being passive. Nutrition advocates rallied publicly against cuts in SNAP, while conservative advocacy groups aired advertisements urging Republican conferees to hold the line. Environmentalists and groups aligned with the National Sustainable Agriculture Coalition urged Senate conferees to defend conservation provisions and budgets. But the biggest fights were old-fashioned public spats between rival commodity groups. The National Corn Growers Association and the American Soybean Association together squared off against groups representing southern rice and cotton, each side issuing dueling broadsides over the specifics of crop insurance versus countercyclical payments and the arcane but fiscally critical issue of basing subsidies on acres owned versus acres planted.[18] The National Cattlemen's Beef Association railed against COOL, which the Ranchers-Cattlemen Action Legal Fund defended. They all worried that conferees eager to reach a deal might settle for far less than these organized groups could stomach. Threats to oppose a final deal became commonplace and changed nearly daily as competing groups intuited momentary shifts in legislative language, often based on little more than whispers from staff members who may or may not have been in the room.[19]

The clear breakdown in the old Farm Bill coalition and their public sniping prompted a rebuke from Peterson. "This is not helping to get a farm bill," he said. "I didn't have this in 2008. The attitude among the commodity groups this time seems to be: line up and shoot." For his part, Lucas saw the big corn and soybean groups as being particularly unwilling to compromise. "For some folks to believe they don't have to be part of the family anymore makes it a little difficult," he said. "As chairman, I'm kind of like a parent sitting at the table. I'm trying to make sure everybody gets their fair portion as the plates go around. I'm trying to make sure the biggest kid doesn't shove all the little kids off the bench."[20]

By late November, it was apparent that the four leaders were bogged down in these various fights and would not meet their self-imposed Thanksgiving deadline. They returned after the holiday even as the Senate remained in recess. Suggestions were floated that Congress pass a simple two-year extension of the 2008 act, but these were quickly shot down as various observers fretted that the conferees might find themselves entangled in yet another round of budget and debt ceiling debates, not to mention another "dairy cliff."[21] That one such suggestion came from the National Corn Growers Association was interpreted as a swipe at Lucas, who would "time out" as House Agriculture Committee chair in 2015 should the Republicans retain control of the House. One representative of the association expressed shock at the suggestion that his group might prefer to work with a new committee chair. "We felt it was important for the conferees to know our position," he explained. "We in no way meant it as a threat."[22] Few believed him.

The conferees reached apparent agreement on two critical issues after returning from their Thanksgiving break. First, Lucas and Peterson gave ground on their chamber's position that revenue and price loss protection programs in the commodity title should be based entirely on average planted acres, a standard favored by Lucas because it reflected farmers' actual exposure. Instead, they agreed to a standard using a percentage of owned or "base" acres, which was favored by midwestern Republican senators such as Pat Roberts, as well as by soy and corn producers who worried that the "planted acres" rule might run afoul of WTO restrictions on direct subsidies. The precise formula would depend on how the CBO scored the various alternatives.[23] Still to be hammered out were differences over annual income thresholds and what it meant to be "actively engaged" in farming to qualify for subsidies, where party mattered less than the individual conferees' beliefs about assistance going to those who were most in need.[24]

The other significant compromise came on SNAP. Once the four principal conferees went into closed discussions, it became clear that they would never approve all the program restrictions and the $39 billion in SNAP cuts contained in the House bill. For one thing, Cochran, the lead Senate Republican negotiator, did not see eye to eye with his House colleagues—SNAP was too important to constituents in his state—and signaled that he was more comfortable with a figure closer to $8 billion. In this regard, he was in sync with Stabenow, who kept reminding her colleagues about the $11 billion in reductions that accompanied the end of the stimulus package enhancements. Stabenow also knew that, regardless of Obama's public opposition to further cuts, Democrats would have to agree to some tightening in program requirements to give Peterson leverage in the House. She proposed to double the Senate bill's SNAP savings to $8 billion by increasing the state LIHEAP threshold to $20 a year.[25] Stabenow also was willing to entertain proposals that might draw some House conferees closer, even at the risk of alienating some Democrats. Notably, one such proposal was aimed at getting Southerland's support by funding pilot projects designed to help jobless SNAP recipients get back into the workforce. Equally if not more important, House Republican conferees, in the words of one, understood that the $39 billion figure in their bill "was never real. It was an exercise we had to go through to get to conference. There was never a real conversation about it in conference."[26]

In mid-December, with Congress scheduled to recess for the holidays, House leaders again floated the idea of a short-term extension of the 2008 law to give conferees a bit more time—and to avoid headlines about Congress failing to act while commodity prices dropped and another "dairy cliff" drew near.[27] The lead conferees, advised by the USDA that milk prices would not spike if a new Farm Bill were in place by early January, were cool to a short-term extension. They were making good progress and wanted to keep the pressure on. "We are very confident that we are going to have an agreement," said Stabenow. "We will be ready to vote in January." Also opposed to an extension was Reid, who was in no mood to take Republicans off the hook just as the 2014 election cycle kicked into gear.[28]

Even so, Lucas hedged his bets, filing for an extension through January 31, 2014.[29] Still to be settled was the ongoing fight over the dairy program, and neither Boehner nor Peterson appeared to be giving much ground. Peterson, though publicly opposed to the extension, did not fight it, and on December 12 the House approved the extension by a voice vote. Reid announced that the Senate would not follow suit, a decision that elicited little criticism from Senate Republicans, who also were not inclined to let their House counterparts off the hook.[30]

That same day the House approved the Bipartisan Budget Act of 2013, which set spending caps for fiscal years 2014 and 2015, raised some additional revenue to be spread among defense and nondefense discretionary spending, and extended sequestration through fiscal year 2023.[31] Of note, House passage (332–94) was as bipartisan as the act's name, suggesting some goodwill as members prepared to adjourn for the holidays. That said, the final vote reflected ongoing tensions among Republicans, with Boehner and Cantor voting yes, and Huelskamp and a core group of conservatives voting no. The House then began its holiday recess.

Meanwhile, the lead conferees continued to talk. They were getting close, Stabenow said, adding that the CBO scoring on commodity programs was looking positive.

Home Stretch—or Brick Wall?

Congress returned from the holiday recess in early January. In public, conferees continued to express optimism. "We're just tying up loose ends," Stabenow cheerfully told reporters. "Feeling very good about things."[32] Behind closed doors, however, skirmishing between Boehner and Peterson over the dairy program had reached a breaking point, with the Speaker personally telling Lucas and Peterson that there would be *no* form of supply management in the final Farm Bill—period. Peterson was equally adamant about retaining some form of production management, despite losing to Boehner on the House floor, and he was increasingly confident of having sufficient votes in conference to override the Speaker. Peterson did not hide his anger: "We're about 12 hours away from not having a farm bill at all if this keeps up," he said.[33] Lucas was caught in the middle. "I don't know that I understood just how hard the positions were by the two interested parties," Lucas said. "No one has shown any flexibility whatsoever." When asked about the matter, Boehner simply responded, "I am confident that the conference report will not include supply and management provisions for the dairy program."[34] Peterson retorted that he, along with Senate Democrats Stabenow and Patrick Leahy of Vermont, had already tried to accommodate Boehner by making the management program voluntary and more narrowly designed to avoid disrupting dairy markets, and by offering to let its stabilization fund expire after three years. "We already compromised three times on dairy. He's compromised zero times," said Peterson. "So that's the problem."[35] Some observers wondered whether Boehner would really block a conference agreement over dairy when he had little to say about similar provisions for sugar, or perhaps his intransigence was a ploy to gain more concessions on SNAP. The Speaker wasn't saying, and supporters of dairy management scrambled to come up with a way to support small producers without incurring his opposition.

The tussle over dairy was the most important of a small and, to the uninitiated, arcane set of conflicts that were still unsettled, including the proposed shift in imported catfish inspection from the FDA to the USDA, the definition of being "actively engaged" in farming, COOL, and the King amendment on state animal welfare laws. Together, however, they were enough to stall movement on the bill, and Lucas admitted that the conference report might not get to the respective chamber floors by the end of the month. In the meantime, Congress managed to remove one potential distraction by agreeing to a fiscal year 2014 appropriations bill. The Farm Bill would hold center stage until mid-February, when Congress would have another go at the debt ceiling, if only Lucas could get a conference report to the floor.

Agreement

The four lead conferees and their staffs continued to meet through January, even during the long Martin Luther King Day weekend, scrambling to work out the small compromises on the remaining vexing issues that might make the difference. Finally, on Monday, January 27, Lucas convened the full committee, which approved the conference report for H.R. 2642, now simply the Agricultural Act of 2014. Even the bill's title had been a point of contention.

Lucas and Stabenow expressed relief. The deal "puts us on the verge of enacting a five-year farm bill that saves taxpayers billions, eliminates unnecessary subsidies, creates a more effective farm safety net and helps farmers and businesses create jobs," Stabenow said. Added Lucas, "I am proud of our efforts to finish a farm bill conference report with significant savings and reforms."[36]

The road to agreement lay in a deal on the dairy program. Forced by Boehner to go back to the drawing board, supporters of dairy management were able to come up with a complicated new insurance subsidy program that addressed some of the needs of smaller producers and, more important, met with Boehner's approval. Critics argued that the new insurance program might prove more expensive in the long run than the USDA-administered production management plan Boehner had opposed, but the long run was a concern for later. Peterson was relieved. "Compromise is rare in Washington these days, but it's what is needed to actually get things done," he said. "My reservations are outweighed by the need to provide long-term certainty for agriculture and nutrition programs. This process has been going on far too long; I urge my colleagues to support this bill and the president to quickly sign it into law."[37]

As befitting such a compromise, the conference agreement offered something for everyone to dislike. On SNAP, conferees agreed to Stabenow's proposal for roughly $8.6 billion in savings, a tightening of program rules, an increase in the minimum annual LIHEAP threshold to $20, additional funds for food banks to address emergency needs, and, based on ideas proposed by Florida's Southerland, pilot projects to help recipients get back into the workforce. "As of right now, we're progressing nicely," said Southerland.[38] Coincidentally or not, the conference report also contained provisions that helped the Florida citrus industry.[39] The SNAP cuts infuriated some nutrition advocates. "They are gutting a program to provide food for hungry people to pay for corporate welfare," said Joel Berg of the New York City Coalition against Hunger.[40] Others were more charitable, calling the outcome "relatively favorable" and certainly better than taking a chance with another Congress.[41] Whether the SNAP reforms were enough to appease a sufficient number of House Republicans remained uncertain.

On commodities, the conference split the difference between House and Senate versions, giving growers a choice between the revenue protection plan in the Senate bill and the countercyclical program promoted by Lucas and the House. The conference version included the House bill's transition program for cotton, modest caps on the annual subsidy amounts available to individuals and married couples ($125,000 and $250,000, respectively), and the Senate bill's requirements that farmers getting subsidized coverage abide by a number of conservation provisions, including reduced subsidies if they plowed up previously pristine prairie lands. Finally, the conference decided to leave the definition of what it meant to be "actively engaged" in farming up to the USDA, except for stating that farmers who owned their land automatically qualified.

Commodity groups declared a truce over the outcome. "The bill is a compromise," said Ray Gaesser, president of the American Soybean Association. "It ensures the continued success of American agriculture, and we encourage both the House and the Senate to pass it quickly."[42] Food system reformers were less charitable, calling the move from direct payments to subsidized shallow-loss insurance a "bait-and-switch" that would ultimately cost taxpayers just as much, if not more, in future years. "This bill is so bad, they literally stripped reform from the title," said Steve Ellis of Taxpayers for Common Sense.[43] Senator Charles Grassley of Iowa, who had long advocated for stricter limits on who was eligible for subsidies, was blunt in his disappointment over the conference's compromise. "This is an example of why Congress has a 12 percent approval rating," he said.[44]

However, real trouble remained as conferees overrode Lucas and voted against repealing the 2002 law's "country of origin" provisions and against supporting language in the House bill overriding USDA regulations on large stockyards, both priorities of the beef industry in particular. Although the conference report called for the USDA to study COOL's economic impacts, the overall outcome infuriated industry groups, which aimed their anger at Stabenow. "We're opposed to the bill, and Debbie Stabenow is to blame. She's the one who said no," said Colin Woodall of the National Cattlemen's Beef Association. When asked if his association would work to take down the entire bill despite appeals by Lucas, who had made sure the bill contained nearly $5 billion in new disaster aid for the livestock industry, Woodall simply said, "We're going to work it hard."[45] Whether they would have much success depended on how much western legislators in particular were willing to side with the meat industry, a consideration conferees had taken into account by adding about $400 million in "payment in lieu of taxes" benefits for rural towns and counties surrounded by nontaxable federal lands.[46]

Finally, conferees kept the new USDA catfish inspection program, disappointing the Senate, and they omitted the King amendment, pleasing animal welfare advocates. Producers of organic fruits and vegetables, no doubt aided by Stabenow, got improved insurance coverage and other production and marketing supports. Of note, the conference bill included some research funds for industrial hemp, so Mitch McConnell got something for his troubles. And, despite keeping intact the longtime international aid program, conferees added another $80 million to enable the US Agency for International Development to purchase food closer to areas in need.[47]

Most telling of all, the conference deleted language in the House bill repealing the 1938 and 1949 commodity titles. The permanent law remained in place. The vestigial farm bloc was not about to give up its nuclear option—

perhaps the only weapon it had left, with the possible exception of a constitutional arrangement whereby the 740,000 residents of North Dakota get as much Senate representation as the 39 million residents of California.

Final Passage

House action on H.R. 2642 was set for Wednesday, January 29. Nancy Pelosi and Collin Peterson urged passage at a January 28 meeting of the Democratic caucus. Equally important, Marcia Fudge, chair of the Congressional Black Caucus, defended the deal in the face of criticism about cuts in food stamps from colleagues such as James McGovern.[48] On the Republican side, John Boehner threw his weight behind the conference report. "The status quo is simply unacceptable," Boehner said. "This legislation . . . is worthy of the House's support."[49] He also promised to vote for it, if necessary, despite never having voted for a Farm Bill in his entire career. How many other members of his caucus would go along was uncertain.

Final House action on H.R. 2642 came the next day, following a last-minute procedural skirmish in which Republicans blunted an effort by Democrats to dedicate a portion of the $23 billion in promised savings to extending unemployment benefits to the long-term jobless.[50] Frank Lucas, given the last word before the vote, asked his colleagues to support the compromise. "Whatever your feelings might be about the policy issues involved within the bill, understand, this formal conference that's now come to a conclusion . . . reflects . . . how legislation should be put together," he said.[51] Observers noted the presence of Debbie Stabenow on the House floor, armed with a list of 400 groups backing the bill to bolster Democratic ranks.[52]

Following a short period of debate, the chamber voted 251–166 in favor of final passage, with 14 not voting. To observers, it was a refreshingly large and bipartisan margin.[53] John Boehner did not cast a vote (returning to the norm of Speakers voting only to break a tie), but Eric Cantor, Kevin McCarthy, and most members of the House Republican leadership supported the bill's passage. So did Budget Committee chair Paul Ryan, despite his complaints that the bill did not go far enough to cut spending or to reform subsidy programs favoring larger producers.[54] Pelosi and other members of the Democratic leadership team supported the deal.[55]

Defections were telling, and they came from both ends of the ideological spectrum. The 103 Democrats voting no were largely the House's most urban and liberal members, and they did so because of cuts in food stamps. "This bill will make hunger worse in America, not better," said McGovern before the vote. "If this bill passes, thousands and thousands of low-income Americans will see their already meager food benefit shrink."[56] Conversely, eighty-nine Democrats supported the conference report, compared with the twenty-four who had voted for H.R. 1947 the previous June. They included half of the forty House members of the Congressional Black Caucus, for whom the "farm programs + food stamps" deal still mattered.[57] Equally important, Peterson regained the votes of rural Democrats who had abandoned him after the defeat of the dairy program in 2013. Every Democratic conferee from the House Agriculture Committee except for McGovern voted in support. Two other Democratic conferees, Eliot Engel and Sander Levin, both members of the Foreign Affairs Committee, opposed the report because it did not contain desired reforms in international aid.

The sixty-three Republican defectors, including all four members from Kansas, were from the party's most conservative wing, and nearly all of them had voted against H.R. 1947 the previous June. As before, they believed the final compromise did not save enough money—only $23 billion over ten years, and possibly less if CBO estimates were overly optimistic. For Tim Huelskamp, the bill did nothing to reform food stamps. "This program is in desperate need of reform, and yet this bill makes only nominal changes," he said.[58] Others conservatives, such as Marlin Stutzman (R-IN), opposed the renewal of a "farm programs + food stamps" linkage, viewing it as "just more business as usual" and noting that it "reverses the victory for common sense that taxpayers won last year. This logrolling prevents the long-term reforms that both farm programs and food stamps deserve."[59] Ed Royce of California, chair of the Foreign Affairs Committee, was the only Republican House conferee and a rare member of the leadership team to vote no, because the bill did not do enough to cut spending or reform international food aid. Despite his amendment's omission from the final package, Steve King of Iowa voted for the bill, as did most rural Republicans. It was an election year.

Hoping to provide momentum prior to the Senate vote, President Obama signaled that he would sign the bill as "currently designed."[60]

On February 3 the Senate approved the conference report on H.R. 2642 by a bipartisan 68–32 vote. The nine Democrats who voted no, largely from the

Northeast, opposed further cuts in SNAP. Republicans were split. Stabenow lost the support of midwestern Republicans such as Charles Grassley and Pat Roberts but made up for it with the votes of southerners such as Thad Cochran, Saxby Chambliss, and, notably, Mitch McConnell. "We like hemp," said a McConnell aide afterward.[61]

Signing Day

On February 7, 2014, President Obama, accompanied by Secretary of Agriculture Vilsack, Senator Stabenow, and other congressional Democrats, traveled to a horse barn at Michigan State University (MSU) in East Lansing to sign H.R. 2642 into law. MSU was the nation's first land-grant agricultural school and, of course, Stabenow's alma mater. In fact, it was Stabenow's idea to sign the Agriculture Act at MSU and not at the White House. "The president needs to sign a farm bill outside of Washington D.C. I'm thrilled he wants to go to America's first land grant university," she said. "I think it's wonderful."[62]

The president, speaking to an audience that included some 500 farmers and local officials, praised "bipartisan" passage of the Agricultural Act, calling it "commonsense reform" that helps small farmers and "lifts up our rural communities" while giving more Americans "a shot at opportunity" in the years ahead.[63] But, he said, "despite its name, the farm bill is not just about helping farmers. Secretary Vilsack calls it a jobs bill, an innovation bill, an infrastructure bill, a research bill, a conservation bill. It's like a Swiss Army knife."[64] Referring to a young farmer present at the signing, the president also made a point to reassert the long-standing connection between federal nutrition programs and a healthy farm economy:

> That's why my position has always been that any farm bill I sign must include protections for vulnerable Americans, and thanks to the good work of Debbie [Stabenow] and others, this bill does that. And by giving Americans more bang for their buck at places like farmers markets, we're making it easier for working families to eat healthy foods and we're supporting farmers like Ben who make their living growing it.[65]

As is traditional, the president made sure to hand out praise—and pens—to the bill's architects. "We had Rep. Frank Lucas, a Republican working with

Collin Peterson, a Democrat," he noted; likewise, Stabenow "had worked with Republican Sen. Thad Cochran, who I think was very constructive in this process."[66] However, only Stabenow was present to hear the president's praise and receive a pen. Lucas and Cochran, both of whom released statements expressing pride in their work on the bill, cited other commitments. Observers noted that both were facing primary challenges from Tea Party–affiliated Republicans and were likely reluctant to be photographed with a president who was not exactly popular with their party's voting base. Peterson also stayed away, despite having worked so hard to get the bill to the president's desk; his absence was taken as a sign that he would run for reelection in his generally conservative district.[67]

In fact, notable by their absence, despite being invited, were any Republicans at all. It had not been your typical Farm Bill. Or had it?

10

What Just Happened Here?

Finding Meaning in the Politics of the Farm Bill

It's never too early to start on the next farm bill.
—Rep. Mike Conaway

At the time of this writing (August 2016), we were just past the midpoint of the authorization of the Agricultural Act of 2014, which will expire on September 30, 2018. Whatever else critics say about this edition of the Farm Bill, it was a significant change from the central thrust of US agricultural policy going back to at least the 1970s. Most notable was the general, if incomplete, shift in the commodity title away from most forms of direct payments and toward an insurance-based "risk management" model. It was seen as a major reform, even if it was accompanied by criticisms that its programs are not well designed, may prove to be more expensive than the system they replaced, and only perpetuate what has become a form of taxpayer-supported welfare for a comparatively small group of affluent farmers.[1]

Other aspects of the 2014 law are worth considering. Of course, most of the attention was focused on the battle over SNAP, and program defenders could claim a general victory, given the extent of threatened eligibility restrictions and funding cuts. However, the outcome in 2014 also meant more overall support for organic and specialty crops, more promotion of local foods and healthy eating, some help for dairy, and a bit less support (at least proportionally) for major commodities.[2] Much to Mitch McConnell's satisfaction, there was even money for research into industrial hemp. In the view of its critics, Congress seemed to give a little something to everyone, with legislators carrying on the tradition of stapling together farm and food interests to get the votes they needed.

In some ways, and all the noise aside, passage of the Agricultural Act of 2014 resembled the normal Farm Bill process. It was messier than usual, to be sure, but Congress ultimately got the job done, and in the end, it did so with bipartisan majorities. So maybe there's no story here. Maybe, as Adler and Wilkerson argue in their comprehensive assessment of congressional problem solving, Congress was confronted by a piece of legislation that showed up, unbidden, on its agenda; wrestled with it a bit (OK, a lot); but ultimately renewed it for a specified number of years, like so many of the "compulsory" duties it fulfills each session.[3] Despite its poor public image, Congress does work. There's nothing to see here. Move along.

Yet, I wonder. After all, the *politics* of the 2014 Farm Bill was especially divisive, and at times it all threatened to fall apart. Why? Was it simply because of the acrid ideological battle over SNAP? No, it was more than that. In pondering that question, it is useful to repeat Nadine Lehrer's explanation for passage of the 1996 farm bill, the previous instance of a major change in US agricultural policy:

> First, Republican control of Congress created an atmosphere in which legislators were looking to limit government intervention in agriculture. Second, there was pressure to reduce the growing budget deficit, for example by reducing commodity subsidies. Third, the General Agreement on Trade and Tariffs [GATT] had highlighted an ideal of liberalized trade to be achieved by countries reducing domestic subsidies and tariffs. Fourth, House Speaker Newt Gingrich (R-GA) authorized commodity programs to be written by budget committees rather than in the more status-quo-oriented agricultural committees. Fifth, the writing of the 1996 bill coincided with a burst of high commodity crop prices in 1995–6.[4]

The political dynamics of 2011–2014 were eerily similar. Although the World Trade Organization has since replaced GATT, the continuation of its free-market trade regime has imposed similar boundaries on domestic agricultural policymaking. Except for programs directed at safeguarding ecologically sensitive lands, the production management techniques and direct subsidies typical of New Deal–era farm policy are, for the most part, forbidden. As highlighted by the adjustments made to the cotton program to respond to a WTO ruling in favor of Brazil and the spat over country-of-origin labeling of meat, the prevailing trade regime was a fact of life for Congress throughout its deliberations—always there, always limiting the range of action, and usually without a lot of talk about it.

Of greater immediate importance to our story, commodity prices during 2011–2013 were at their highest in years—the highest since 1995, in fact—and farmers overall were doing better than at any time in memory. For many in the all-important Midwest Grain Belt, it was agriculture's new "golden age." As a result, most commodity producers were not banging on Washington's door and demanding help. In fact, as congressional Democrats found out, rural America was feeling flush enough to vote based on cultural values, not economic need. Unlike the late 1970s or mid-1980s, there were no visible crises in the fields, no Washington "tractorcade" of bankrupt farmers, no nationally televised Farm Aid concerts by Willie Nelson and John Mellencamp.[5] Although many small dairy farmers in New England and the Midwest still struggled, and although natural disasters like the upper Plains drought still had localized impacts, a generally prosperous farm sector looked on the congressional debates over commodity policy with little interest or intensity. In fact, farmers seemed to keep a low profile throughout. In their absence, others defined the contours of Farm Bill debate.

For the most part, those others were the most ideologically conservative members of the Republican Party. As in 1995, the Republican takeover of the House in 2011 would define the lengthy battle over the Agricultural Act of 2014, but not in a narrowly partisan way. Even including the partisan divisions over SNAP, the most telling fights over the 2014 act were *among Republicans.* By 2011, there were so few rural conservative House Democrats that Collin Peterson's cage match with John Boehner over the dairy program was the exception, not the rule. More typical were tensions between older, socially conservative, but generally pragmatic Farm Belt Republicans like Frank Lucas and Pat Roberts and newer, more doctrinally conservative Republicans like Tim Huelskamp and Marlin Stutzman. By 2011, Huelskamp, despite his own Kansas farm roots, was more in sync with the new conservative majority than was cattle rancher Lucas. Throughout, the members of a dwindling traditional farm bloc were caught in an intraparty fight between the economic self-interests of a small and comparatively well-off group of commodity producers and the interests of a core nonfarm Republican voting electorate demanding a smaller and less intrusive government, at least in the abstract. Given that power in the House resides in the hands of the majority, those intraparty cleavages defined Farm Bill politics and outcomes. In the Senate, over which Democrats maintained control during this period, most of the fights reflected regional commodity demands, with midwestern defenders of corn and soybeans pitted against southern defenders of rice and cotton. But those fights also occurred mostly among Republicans—Roberts versus Cochran, for example—as Senate Democrats more typically come from coastal, Great Lakes, and upper Plains states with far different farm and food policy interests.

Of greater importance, as it had been in 1995–1996, was Republican dominance over the broader congressional agenda, especially in the House. Nowhere was this clearer than in the Republicans' zeal to reduce a federal budget deficit largely created, their critics were quick to point out, during a previous Republican presidency. No matter: Barack Obama's election and actions in his first two years energized the conservative base, to the benefit of congressional Republicans. With its renewed control of the House in January 2011, the overriding goal for the Republican caucus was to slash nondefense spending and get the federal government out of the business of regulating the economy, including agriculture. As a result, the politics of the Agricultural Act of 2014 took place within the context of budget cuts and demands for deep policy change. Politically indefensible direct payments were out, and Lucas had to start with the reality of less money for commodity programs. It was no wonder that he and fellow farm bloc legislators looked longingly for savings in SNAP, which now accounted for 80 percent of annual Farm Bill spending. Given the continuation of the budget sequester—which each year automatically takes yet another chunk out of discretionary spending and, in fact, has helped lower the federal deficit—such pressures will only continue when the Farm Bill next comes up for renewal.

As in 1995–1996, the centrality of the budget in 2011–2014 momentarily shifted power in Congress away from the traditional defenders of agriculture. If John Boehner did not actually give Paul Ryan direct control over the House bill, Ryan and the Budget Committee effectively dictated what Lucas and the Agriculture Committee could spend, knowing that Boehner would make those numbers stick. The problem for Boehner, of course, was that no matter how much Lucas cut, it was never enough for the Tea Party conservatives whose election had brought Boehner to power in the first place. That Boehner could block action on the Farm Bill through 2012 with no damage to his leadership, or that he could threaten to torpedo the eventual conference package over the offending dairy management program, underscored agriculture's diminished clout in the House. It says a lot when what's left of the congressional farm bloc hinges its hopes on the Senate's structural bias favoring rural America. Frank Lucas and Collin Peterson should send a thank-you card to North and South Dakota.

In the end, and after considerable torture and seemingly endless budget cliffhangers that threatened to suck the energy out of the entire legislative process, the farm bloc managed to eke out a win. And, irony of ironies, success may have come largely because of majority leader Eric Cantor. For reasons of his own, Cantor pulled SNAP out of the House bill after the disastrous floor defeat of June 2013, got stand-alone SNAP and "farm only" bills approved by straight party-line votes, and then joined the two back together before shipping the House package to conference. The "farm programs + food stamps" coalition, thus reestablished, ultimately held. But it was a close call, closer than promoters of agriculture cared to admit.

Consequences

According to Steffen Schmidt of Iowa State University, "The farm bill is a decennial 'seasonal' political event. Like the coming of locusts it appears, is loud and seems hugely important and then subsides and vanishes."[6] Given its timing, did the battle over the Agricultural Act of 2014 have any effect on the November 2014 elections? For the most part, no—at least not directly. Despite lingering bitterness among farm and commodity groups about House Republicans' mishandling of their farm bill through 2013 and general criticism about their performance overall, Republicans actually did very well in what turned out to be another classic midterm election in which energized conservative voters—most of them white, older, and comparatively well-off—again made the difference in what was otherwise the lowest voter turnout (36.3 percent) since 1942.[7] As in 2010, conservatives managed to make the midterm elections all about Barack Obama, whose 2012 electoral coalition simply stayed home. As a result, Republicans picked up an additional thirteen House seats, most at the expense of moderate to conservative Democrats in rural districts—including four who sat on the House Agriculture Committee—and started the 114th Congress with their largest majority since 1928. John Boehner returned as Speaker. Republicans also picked up nine Senate seats, largely in the Midwest and South, and regained control for the first time since 2007. Mitch McConnell and Harry Reid switched positions when the 114th Congress convened in January 2015.

For some Republicans, in fact, their *opposition* to the Farm Bill proved pivotal. In Arkansas, Tom Cotton, first elected to the House in 2012, defeated incumbent Democratic senator Mark Pryor, despite Cotton's vote against the conference compromise and Pryor's dogged support for the state's rice and catfish sectors. In fact, Pryor had been instrumental in defending the USDA catfish inspection station that critics saw as redundant and protectionist—all the better for Arkansas aquaculture. Yet Cotton adroitly deflected Pryor's attacks on his nay vote as a "slap in the face to farmers" by turning the issue on its head: his vote, he said, was against *Barack Obama.* "When President Obama hijacked the farm bill and turned it into a food stamp bill with billions more in spending, I voted no," Cotton said in a statewide campaign ad.[8] Despite their state's reliance on agriculture and comparatively high rate of SNAP utilization, the 41.2 percent of Arkansas voters who showed up on Election Day disliked Obama even more, and Cotton won easily. The same dynamic played out in Iowa, where Tea Party–backed Joni Ernst harvested conservative votes on issues such as same-sex marriage to replace retiring Democrat Tom Harkin—and despite her promise to kill the federal renewable fuel standard that heavily subsidized the state's corn ethanol industry.[9] Nowhere, it seemed, did voting against the Farm Bill matter.

The few casualties in 2014 were suggestive. Steve Southerland, author of the SNAP floor amendment that had blown up H.R. 1947 in June 2013, lost a close race in a somewhat competitive district to Democrat Gwen Graham, whose name recognition was aided by being the daughter of a former Florida governor.[10] Frank Lucas campaigned on his behalf among farmers, but Southerland committed some highly publicized campaign gaffes that likely contributed to his loss. Notably, he hosted an expensive "men only" fund-raising dinner that Graham's campaign contrasted unfavorably with Southerland's advocacy of tougher work requirements for SNAP recipients, the majority of them women with children.[11] Southerland subsequently joined a Capitol Hill consulting firm. Graham would serve only one term, deciding not to seek reelection in 2016 after a court-ordered statewide redistricting plan caused African American voters to be shifted to an adjacent district, leaving hers a more secure Republican seat than it had been.[12]

In perhaps the supreme irony, the most visible loser in 2014 was House majority leader Eric Cantor, "savior" of the House Farm Bill. He suffered a shocking—to everyone—primary defeat to a libertarian conservative economics professor whose ideology ran to Cantor's right. Congressional leaders rarely lose reelection—Cantor may well have been the first House majority leader *in history* to lose in a party primary—because those who rise to leadership positions generally represent safe seats whose voters afford them the luxury of focusing on internal House matters. But Cantor apparently forgot former Speaker Tip O'Neill's cardinal rule: you still need to go out and ask for votes.[13] Cantor apparently didn't do it enough, whereas his opponent campaigned hard and got significant airplay from influential conservative radio talk-show hosts who felt that Cantor and other Republican leaders simply had not delivered on their promises. Despite heavy last-minute advertising, Cantor lost by a convincing ten percentage points. Adding to the irony was that the Republican-dominated Virginia legislature had recently redrawn Cantor's district to include *more Republican voters*, presumably to make it safer for him.[14] It did just the opposite, underscoring the ideological war among Republicans about who is conservative enough, which continued rather spectacularly in their 2016 presidential primary.

Cantor and Southerland aside, the central players in the saga of the 2014 Agricultural Act survived and returned to Congress in 2015. Frank Lucas easily beat back a primary challenge from his right and went around the country campaigning on behalf of fellow Republicans—even some who had voted against the Farm Bill. Because of Republican caucus term limits on committee chairs, Lucas was replaced in that role by Michael Conaway of Texas. Should Republicans retain the House after the 2016 elections, Conaway will lead the House Agriculture Committee when the Farm Bill comes up for renewal in 2017–2018. Collin Peterson, despite speculation that he might retire, ran again and won what turned out to be a rather hotly contested race in his otherwise conservative rural district. He returned in 2015 as ranking minority member of the Agriculture Committee and has indicated his intent to seek reelection in 2016, presumably to be present when it's time to do it all again.

Pat Roberts survived a tough challenge from a Tea Party–endorsed candidate, winning his primary by only 7 percent. Roberts campaigned on pledges to cut SNAP spending and rein in the Environmental Protection Agency's authority over agricultural practices, but his votes on the 2014 act were almost a sideshow compared with the controversy over whether he actually owned a home in the state he had represented since 1980.[15] Even so, Roberts went on to a comfortable general election win in his heavily Republican state.[16] With Republicans in control of the Senate, Roberts in 2015 achieved his goal of chairing the Committee on Agriculture, Nutrition, and Forestry; Thad Cochran opted to chair the Appropriations Committee instead. Despite concerns among farm and food groups that Debbie Stabenow would leave the Agriculture Committee to be the ranking minority member on the Budget Committee, she decided to stay on.[17] Stabenow and Roberts will likely play key roles in Farm Bill deliberation in the 115th Congress, but those roles will depend on whether Republicans can hold on to the Senate.

House Republicans added seats to their majority after the 2014 elections, but that didn't help John Boehner. In fact, it made his life worse. In a surprise move, he resigned as Speaker in October 2015, citing fatigue caused by the incessant battles with his party's most conservative bloc, now known as the Freedom Caucus. He was only the second House Speaker in 150 years to resign in the middle of a term, the other being James Wright in 1989 following allegations of ethical misconduct. Notably, Boehner was the *only* House Speaker in US history to resign because of internal schisms in his majority caucus.[18] After considerable uncertainty and amidst criticism about their purported failures at governance, House Republicans eventually convinced Budget Committee chair Paul Ryan to take up the gavel. To date, Ryan has fared a bit better with his most conservative colleagues, although he too has struggled to forge consensus among Republicans on budget and appropriations bills.[19]

As for Tim Huelskamp, his vocal opposition to the Farm Bill, his fights with Boehner that led to his ouster from the House Agriculture Committee, and, as if to accentuate his maverick stance, his 2014 election-year cosponsorship of a bill to phase out the renewable fuel standard infuriated Kansas agricultural interests and earned him a comparatively well-financed primary challenge by Alan LaPolice, a farmer and community college teacher. Huelskamp was criticized throughout the primary campaign by the proverbial

Kansas agricultural establishment, including the Kansas Corn Growers Association, Kansas Farm Bureau, Kansas Association of Ethanol Processors, and Kansas Grain Sorghum Producers Association. Equally telling, he was not endorsed by his former ally, the Kansas Livestock Association.[20] No matter: backed by the funds of national conservative advocacy groups such as FreedomWorks and the Tea Party Express, Huelskamp won the primary with 55 percent of the vote and went on to an easy victory in the general election. So did the other three members of the Kansas congressional delegation, despite their no votes on the Agricultural Act.

Huelskamp's reelection in 2014 brings us back to the puzzle that prompted this book in the first place: How could he go against the key agricultural interests in his home state and not pay the price at the polls? Weren't they his constituents? Well, maybe not, if demographics mean anything. The Big First may be characterized by farming, but missing from it are *farmers.* In contrast to the middle of the twentieth century, when 25 percent of the district's population worked on farms and did their business in regional centers such as Hays and Garden City, farmers are now less than 2 percent of the district's population. Much of the Big First is really the Big Empty—vast stretches of farmland with few people on it. In fact, most who live and vote in the district reside in small cities such as Manhattan and Emporia or in the outer suburbs ringing Topeka and Wichita, and they make their living in manufacturing and service sector jobs. Agriculture, by contrast, accounts for only 10 percent of the district's employment base, largely consisting of Mexican and other Hispanic immigrants employed in meatpacking—and who don't vote. In 2014 the voters of the Big First, reflecting views held in similar districts throughout the Midwest, South, and Southwest, sent Huelskamp and like-minded conservatives back to Congress to promote their social values, get rid of Obamacare, and rein in federal spending—without touching national defense or "earned" entitlements like Social Security and Medicare.

By 2016, Huelskamp was back inside the tent, insofar as Speaker Ryan had appointed him to serve as an at-large member of the Republican Steering Committee, a key leadership panel that, among other duties, allocates Republican seats on House committees. He still did not have a seat on the House Agriculture Committee, but he promised his district's voters that he would get back on that panel after the 2016 elections. When asked by reporters, neither Speaker Paul Ryan nor Agriculture Committee chair Michael

Conaway could guarantee that Huelskamp would get his wish. Meanwhile, Huelskamp was out on the stump almost every weekend seeking votes at a retail level, most visibly at "town hall" meetings in each of the sixty-three counties that make up the Big First. As of July 4, 2016, he was on number 372.[21]

Coda: Agriculture's Revenge?

That vignette about Huelskamp's peripatetic campaigning was meant to be the end of his part of this story. Although he faced a challenge in the Republican primary election, he was expected to win. Incumbent members of Congress rarely lose, especially in primary elections. However, as is always the case in politics, circumstances can change rather quickly.

By the August 2 primary, it had become clear that voters in the Big First had tired of their pugnacious representative. This time, Huelskamp faced a challenge by Roger Marshall, a fifth-generation farmer, obstetrician, and hospital executive with no previous electoral experience. Marshall, like LaPolice in 2014, made Huelskamp's vote against the Farm Bill and ouster from the Agriculture Committee key campaign issues. In what must have been a particularly ironic twist, Marshall also cited the incumbent's 1995 dissertation—which critiqued the economics of commodity subsidies and portrayed the House and Senate Agriculture Committees as "outlier panels" that "over-represent agricultural and rural interests"—as proof that Huelskamp was an apostate on agriculture and unsuited to continue to represent the Big First.[22] Throughout the campaign, highlighted by rare (for a House primary) multiple debates between the two, Marshall depicted himself as a pragmatist who would support Kansas agriculture first and foremost.[23] He gained the endorsement of the Kansas Livestock Association and Kansas Farm Bureau (former Huelskamp allies) and also won the support of the American Farm Bureau Federation and the National Association of Wheat Growers.[24] Agriculture had picked sides.

Even so, Marshall might not have fared much better than LaPolice in 2014 were it not for outside organizations that backed his challenge with money. While Huelskamp got financial support from National Right to Life, the National Rifle Association, and conservative groups such as Club for Growth and Americans for Prosperity, Marshall got critical backing from what passes these days for establishment Republicanism—notably, the US Chamber of Commerce—and by business-backed advocacy organizations such as Strong Leadership for America and the End Spending Action Fund. Known donors to the latter included Kansas-based ethanol firms that were upset by Huelskamp's support for elimination of the federal renewable fuel standard that props up the biofuels industry. Other funds came from the sugar industry, in response to Huelskamp's efforts to eliminate sugar subsidies.[25] In all, outside groups poured more than $2 million into the race on Marshall's behalf, fueling a barrage of television and radio ads.[26]

That money made all the difference. On August 2, 2016, Huelskamp lost to Marshall by a stunning thirteen points—a result that former Speaker Boehner apparently celebrated with a glass of red wine.[27] Barring a cataclysm in the November general election, Marshall will be the Big First's latest in a long line of Republican representatives. He has pledged to regain the district's rightful place on the House Agriculture Committee. Most observers expect him to get his wish, in pointed contrast to Speaker Ryan's reticence about guaranteeing a seat for Huelskamp.[28]

On one level, Huelskamp's ouster could be interpreted as revenge by the agricultural establishment, which took the rare step of getting deeply involved in a party primary battle. Indeed, despite his hundreds of "town meetings," Huelskamp lost critical votes in the western parts of the district, where farming dominates. But it was more than that. After all, the other three members of the Kansas House delegation won their respective primaries, despite having voted against the Agricultural Act of 2014. Huelskamp lost not because of angry farmers but because his core base in the suburbs outside Wichita and Manhattan simply decided that he had worn out his welcome. Marshall was a credible replacement.

At another level, Huelskamp's loss had more to do with internal Republican Party politics. The primary was the First Kansas version of the Spanish Civil War, with outside powers using local candidates as proxies in a larger conflict. Huelskamp may have put off constituents by being stubborn and opinionated, but he was no different in 2016 than he had been in 2014 or at any other time since first winning a seat in the state senate in 1995. He lost in August 2016 because he faced a credible opponent in a one-on-one race that was massively financed by outsiders motivated by their own agendas. Agriculture got its revenge, but as a side benefit of a broader war for control being fought among Republicans. Huelskamp's allies in the conservative movement have taken names; Speaker Ryan's is at the top of the list.[29]

We'll Meet Again, Don't Know How, Do Know When

The Agricultural Act will come up for renewal in the 115th Congress (2017–2018).[30] As has been the case since 1949, many of its commodity programs will lapse if not reauthorized before the deadline. Despite efforts in the House to repeal agriculture's "permanent law," it remains in place, forcing those who care about agriculture—and the great many more who don't—to go at it once again.

In the meantime, conditions in the field have changed—again. Commodity prices, particularly for corn and soybeans, have slumped badly since their historic highs of 2012. Owing in part to a weakened Chinese economy, global demand for US dairy products has dropped so much that the USDA announced in August 2016 that it would purchase and store 11 million pounds of cheese, which it intends to distribute to the food insecure through food banks.[31] That familiar combination of unlimited production and stagnant demand has led to the lowest farm incomes since 2002.[32] The fiscal impact, as critics foresaw, has been a dramatic increase in crop insurance payouts and the highest commodity program costs since 2006.[33] Put simply, the commodity title of the 2014 act has not saved a dollar. In fact, it is costing billions more than expected, which always seems to be the case when Congress bases projected savings on high commodity prices.[34] But prices always seem to go down, and if they stay low, farmers will demand action to protect their government safety net.

Faced with that likelihood, the House and Senate Agriculture Committees are likely to look for savings in nutrition programs—again. In this regard, the 2014 act's touted $800 million a year in promised SNAP cuts has not panned out. For one thing, most of the proposed savings were supposed to come through a tightening of the "loophole" whereby states were required to boost LIHEAP minimums to $20 a year for SNAP recipients to qualify for higher benefits. But governors in twelve of the sixteen states providing heating assistance decided that spending a few hundred thousand dollars in additional LIHEAP subsidies to meet the $20 threshold was worthwhile if it leveraged tens of millions in additional federal SNAP benefits. Sponsors of LIHEAP "reform" cried foul, but nutrition program defenders argued that the states were acting within the letter of the law to protect their residents.[35]

Meanwhile, SNAP caseloads and annual expenditures continued to fall from their historic highs of 2010–2011, largely due to an improving economy as recipients found jobs or, as critics complained, some states reimposed pre-recession time limits and removed some recipients from eligibility.[36] Whatever the cause, annual SNAP spending fell from $79.8 billion in fiscal year 2013 to $73.9 billion in fiscal year 2015.[37] Even so, program expenditures are still above prerecession levels, and as the largest overall portion of Farm Bill spending, SNAP remains a target. House Republicans in 2015 again passed a budget that turned the program into a set of block grants to the states, but their Senate colleagues refused to go along.

For his part, House Agriculture Committee chair Michael Conaway has used the 114th Congress to hold a series of committee hearings on the program. The problem with the debate over SNAP in 2013, Conaway concluded, was "we went after the money" without a consensus on how to meet the nutritional needs of the food insecure. As such, Conaway has opposed altering SNAP through the annual budget and appropriations process. To date, while he has been noncommittal about maintaining the "farm programs + food stamps" connection when the Farm Bill comes up in the 115th Congress, he has cautioned commodity producers that they need to better educate urban consumers about the importance of agricultural programs to food availability and price, "to make a rural-urban coalition that is not anchored in food stamps."[38]

The question facing Conaway and what remains of the congressional farm bloc is what a rural-urban coalition might look like without nutrition programs. Despite the wishes of ideological conservatives like Huelskamp, those defending that always awkward marriage of convenience are unlikely to risk divorce. Agriculture groups know that urban America, where most of us now live, has little desire to support what amounts to a generous welfare program for a comparatively small group of farmers, regardless of their privileged status in the national cultural narrative, unless something is also done for those who lack ready access to affordable food. The stark reality for Conaway, Lucas, Peterson, Roberts, and Stabenow is that without SNAP, farm programs will be more vulnerable to attack, whether by conservatives who are critical of the free-market distortions of commodity subsidies or environmentalists who condemn the ecological impacts of overproduction.

In short, without that connection to SNAP, defenders of commodity supports may well end up on a rather isolated island, increasingly dependent on the capacity of senators from rural states to hold back the demographic tide. For their part, nutrition program advocates know that without a tight tether to commodity programs—and to agriculture's permanent law—SNAP will end up on its own lonely island, loved by few and much more vulnerable to drastic budget cuts, particularly if Republicans maintain their hold over Congress after 2016. Like two partners in an awkward marriage of convenience, never really in love, farm programs and food stamps still need each other, possibly forever.

What will happen to the Farm Bill in the 115th Congress? Predictions are foolhardy, as the outcome depends on the yet-to-be-determined interplay of many factors: which party controls the House and the Senate, and by how much; the relative pain of farmers out in the fields owing to the effects of commodity prices and the costs of production; to what extent the federal budget deficit, which continues to drop from its 2012 high, dominates the larger policy agenda; and who is president—perhaps in that order.

One thing is safe to say: come 2017, defenders of agriculture will have their work cut out for them; the Farm Bill will be back on their—our—agenda, like it or not.

NOTES

CHAPTER ONE. WHAT'S GOING ON IN KANSAS?

For my choice of chapter title, apologies to Thomas Frank, *What's the Matter with Kansas? How Conservatives Won the Heart of America* (New York: Henry Holt, 2004).

1. Census of Agriculture, 2012, *Congressional District Profile*, National Agricultural Statistics Service, US Department of Agriculture, http://www.agcensus.usda.gov/Publications/2012/Online_Resources/Congressional_District_Profiles/cd2001.pdf.

2. See Michael Barone and Chuck McCutcheon, *The Almanac of American Politics 2014* (Chicago: University of Chicago Press, 2013), 673–676.

3. See Jonathon Strong, "Dissidents Pushed off Prominent Committees," *Roll Call*, December 3, 2102, http://www.rollcall.com/news/dissidents_pushed_off_prominent_committees-219624-1.html.

4. Larry Dreiling, "Huelskamp Removed from House Ag Committee," *High Plains Journal*, December 10, 2012.

5. Ron Nixon, "House Approves Farm Bill, Ending a 2-Year Impasse," *New York Times*, January 29, 2014, http://www.nytimes.com/2014/01/30/us/politics/house-approves-farm-bill-ending-2-year-impasse.html?_r=0.

6. David Mayhew, *Congress: The Electoral Connection* (New Haven, CT: Yale University Press, 1974).

7. See Theda Skocpol and Vanessa Williamson, *The Tea Party and the Remaking of Republican Conservatism* (New York: Oxford University Press, 2013).

8. Ron Nixon and Derek Willis, "No Help for Farm Bill from Miffed Kansans in the House," *New York Times*, January 31, 2014, http://thecaucus.blogs.nytimes.com/2014/01/31/no-help-for-farm-bill-from-miffed-kansans-in-the-house/.

9. Based on the composite *National Journal* conservative scale; see Steve Kraske and Dave Helling, "Study: Kansas Congressional Delegation Is Most Conservative," *Kansas City Star*, May 7, 2012, http://www.mcclatchydc.com/2012/05/07/147922_study-kansas-congressional-delegation.html?rh=1.

10. As provocatively detailed in Frank, *What's the Matter with Kansas?*

11. Timothy A. Huelskamp, "Congressional Change: Committees on Agriculture in the U.S. Congress" (Ph.D. diss., American University, 1995).

12. Emma Dumain, "Farm Bill Finally Passes the House," *Roll Call*, January 29, 2014, http://blogs.rollcall.com/218/farm-bill-finally-passes-the-house/?dcz=.

13. Eric Schlosser, *Fast Food Nation: The Dark Side of the All-American Meal* (New York: HarperCollins, 2002); Marion Nestle, *Food Politics: How the Food Industry Influences Nutrition and Health* (Berkeley: University of California Press, 2002); Michael Pollan, *The Omnivore's Dilemma: A Natural History of Four Meals* (New York: Penguin, 2006); Daniel Imhoff, *Food Fight: A Citizen's Guide to the Next Food and Farm Bill* (Healdsburg, CA: Watershed Media, 2012).

14. Upton Sinclair, *The Jungle* (New York: Doubleday, 1906). Although his famous quote has been reprinted in many places, its origin is uncertain, but it might be the October 1906 issue of *Cosmopolitan.*

15. E. E. Schattschneider, *Politics, Pressures, and the Tariff* (New York: Prentice-Hall, 1935); J. Leiper Freeman, *The Political Process* (New York: Random House, 1955); Raymond Bauer, Ithiel de Sola Pool, and Lewis Anthony Dexter, *American Business and Public Policy* (New York: Atherton, 1963); A. Grant McConnell, *Private Power and American Democracy* (New York: Alfred A. Knopf, 1966); Emmette S. Redford, *Democracy in the Administrative State* (New York: Oxford University Press, 1969); Jeffrey M. Berry, *Feeding Hungry People: Rulemaking in the Food Stamp Program* (New Brunswick, NJ: Rutgers University Press, 1984); William P. Browne, *Private Interests, Public Policy, and American Agriculture* (Lawrence: University Press of Kansas, 1988); John Mark Hansen, *Gaining Access: Congress and the Farm Lobby, 1919–1981* (Chicago: University of Chicago Press, 1991).

16. See Adam Sheingate, *The Rise of the Agricultural Welfare State* (Princeton, NJ: Princeton University Press, 2000); Robert Paarlberg, *Food Politics: What Everyone Needs to Know* (New York: Oxford University Press, 2010).

17. Economic Research Service, US Department of Agriculture, *Ag and Food Sectors and the Economy*, May 14, 2015, http://ers.usda.gov/data-products/ag-and-food-statistics-charting-the-essentials/ag-and-food-sectors-and-the-economy.aspx.

18. "203 Rural, 50 Suburban, 103 Urban, 79 Mixed Districts," in *CQ Almanac 1963*, 19th ed. (Washington, DC: Congressional Quarterly, 1964), 1170–1172, http://library.cqpress.com/cqalmanac/cqal63-1315529.

19. "Analyzing Characteristics of the 113 Congressional Districts," 2010 US census data as formulated by Proximity One, http://proximityone.com/cd113_2010_ur.htm.

20. See Bill Bishop, *The Big Sort: Why the Clustering of Like-Minded America Is Tearing Us Apart* (Boston: Houghton Mifflin Harcourt, 2008); Sean Cunningham, *American Politics in the Postwar Sunbelt* (New York: Cambridge University Press, 2014).

21. See Browne, *Private Interests, Public Policy, and American Agriculture.*

22. James Madison, Alexander Hamilton, and John Jay, *The Federalist Papers*, ed. Clinton Rossiter (New York: New American Library, 1961); Robert Dahl, *A Preface to Democratic Theory* (Chicago: University of Chicago Press, 1956).

23. See Hansen, *Gaining Access.*

24. Michael Pollan, "You Are What You Grow," *New York Times Magazine*, April 22, 2007.

25. See, among others, Frank, *What's the Matter with Kansas?*; Skocpol and Williamson, *The Tea Party and the Remaking of Republican Conservatism*; Thomas Schaller, *The Stronghold: How Republicans Captured Congress but Surrendered the White House* (New Haven, CT: Yale University Press, 2015).

26. Michael Shear, "In Signing Farm Bill, Obama Extols Rural Growth," *New York Times*, February 7, 2014, A12.

CHAPTER TWO. THE FOOD SYSTEM: OR, WHY GOVERNMENTS DON'T LEAVE AGRICULTURE TO THE MARKETPLACE

Epigraph: Wendell Berry, "The Pleasures of Eating," in *What Are People For?* (New York: North Point Press, 1990), 145–152.

1. William H. Blair, "Butz Says He'll Speak for All Farmers," *New York Times*, November 18, 1971.

2. Portions of this section are adapted from Christopher Bosso and Nichole Tichenor, "Eating and the Environment: Ecological Implications of Food Production," in *Environmental Policy: New Directions for the 21st Century*, 9th ed., ed. Norman Vig and Michael Kraft (Washington, DC: CQ Press, 2015), 194–214.

3. "A Special Tribute to Earl Butz," *Farm Futures*, February 4, 2008, farmfutures.com/story-a-special-tribute-to-earl-butz-17-28780.

4. USDA, *Food Expenditures*, table 7, www.ers.usda.gov/data-products/food-expenditures.aspx#.U3UFT17rVa.

5. USDA, *Farms and Farm Acreage*, www.nass.usda.gov/index.asp; USDA, *2012 Census of Agriculture*, www.agcensus.usda.gov/Publications/2012/Preliminary_Report/Full_Report.pdf.

6. USDA, *2012 Census of Agriculture*, www.agcensus.usda.gov/Publications/2012/Full_Report/Volume_1,_Chapter_1_US/st99_1_002_002.pdf.

7. Purdue University, 2002, cited in Environmental Protection Agency, *Farm Demographics*, www.epa.gov/oecaagct/ag101/demographics.html.

8. USDA, *U.S. Corn Acreage and Yield*, www.ers.usda.gov/media/521667/corndatatable.htm; USDA, *Background on Corn*, www.ers.usda.gov/topics/crops/corn/background.aspx#.U3Tpr17rVa8.

9. Michael Pollan, *The Omnivore's Dilemma: A Natural History of Four Meals* (New York: Penguin, 2006), 63.

10. American Meat Institute, *The U.S. Meat Industry at a Glance*, www.meatami.com/ht/d/sp/i/47465/pid/47465; USDA, *Livestock and Meat Domestic Data*, www.ers.usda.gov/data-products/livestock-meat-domestic-data.aspx#.U3Yb817rVa8.

11. USDA, *Dairy: Overview*, www.ers.usda.gov/topics/animal-products/dairy/background.aspx#.UzmFzNyE4dI; USDA, *Milk: Production per Cow by Year*, www.nass.usda.gov/Charts_and_Maps/Milk_Production_and_Milk_Cows/cowrates.asp.

12. James McDonald and William McBride, "The Transformation of U.S. Livestock Agriculture: Scale, Efficiency, and Risks," 2009, www.ers.usda.gov/publications/eib-economic-information-bulletin/eib43.aspx#.U3tToV7rVa8.

13. Stuart Melvin et al., "Industry Structure and Trends in Iowa," in *Iowa Concentrated Animal Feeding Operations Air Quality Study*, University of Iowa, 2003, www.public-health.uiowa.edu/ehsrc/CAFOstudy/CAFO_finalChap_2.pdf.

14. USDA, *Hogs & Pork*, www.ers.usda.gov/topics/animal-products/hogs-pork.aspx.

15. USDA, *2012 Census of Agriculture*, www.agcensus.usda.gov/Publications/2012/Full_Report/Volume_1,_Chapter_1_US/st99_1_055_055.pdf.

16. One Btu equals the amount of energy needed to heat or cool one pound of water by one degree Fahrenheit. See Jayson Beckman et al., *Agriculture's Supply and Demand for Energy and Energy Products*, 2013, www.ers.usda.gov/media/1104145/eib112.pdf.

17. Patrick Westhoff, *The Economics of Food: How Feeding the Planet Affects Food Prices* (Upper Saddle River, NJ: Pearson, 2010).

18. USDA, *U.S. Agricultural Trade*, www.ers.usda.gov/topics/international-markets-trade/us-agricultural-trade.aspx#.UzrIU9yE4dI.

19. USDA, *Import Share of Consumption*, www.ers.usda.gov/topics/international-markets-trade/us-agricultural-trade/import-share-of-consumption.aspx#.UzrL8tyE4dJ.

20. Marc Eisner, Jeff Worsham, and Evan Ringquist, *Contemporary Regulatory Policy*, 2nd ed. (Boulder, CO: Lynne Rienner, 2006).

21. Andrew Ross Sorkin, *Too Big to Fail: The Inside Story of How Wall Street and Washington Fought to Save the Financial System—and Themselves* (New York: Viking, 2009).

22. See Bruce L. Gardner, *American Agriculture in the Twentieth Century: How It Flourished and What It Cost* (Cambridge, MA: Harvard University Press, 2002).

23. Stephanie Strom, "Bird Flu Sends Egg Prices up, but Slowing Demand Prevents Shortages," *New York Times*, June 17, 2015, B3, http://www.nytimes.com/2015/06/17/business/bird-flu-sends-egg-prices-up-but-slowing-demand-prevents-shortages.html.

24. Fred Kirschenmann, "A Food and Farm Bill for the 21st Century," in *Food Fight: The Citizen's Guide to the Next Food and Farm Bill*, ed. Daniel Imhoff (Healdsburg, CA: Watershed Media, 2012), 9.

25. Lydia Mulvany, "The U.S. Is Producing a Record Amount of Milk and Dumping the Leftovers," *Bloomberg Business News*, July 2, 2015, http://www.bloomberg.com/news/articles/2015-07-01/milk-spilled-into-manure-pits-as-supplies-overwhelm-u-s-dairies.

26. Earl O. Heady, *A Primer on Food, Agriculture, and Public Policy* (New York: Random House, 1967), 11.

27. Andrew Holland, "The Arab Spring and World Food Prices," *Climate Security Report*, American Security Project, November 2012, http://www.americansecurityproject.org/climate-change-the-arab-spring-and-food-prices/. For a nuanced assessment, see Westhoff, *Economics of Food.*

28. See William P. Browne et al., *Sacred Cows and Hot Potatoes: Agrarian Myths in Agricultural Policy* (Boulder, CO: Westview Press, 1992).

29. To read the 2008 act in its entirety, see http://www.gpo.gov/fdsys/pkg/BILLS-110hr2419enr/pdf/BILLS-110hr2419enr.pdf.

30. See Jennifer Clapp, *Hunger in the Balance: The New Politics of International Food Aid* (Ithaca, NY: Cornell University Press, 2012).

CHAPTER THREE. HISTORY IS NOT BUNK: HOW FARM BILLS PAST SHAPE FARM BILLS PRESENT

Epigraph: Quoted in Jerry Hagstrom, "A Conversation with Frank Lucas," *Hagstrom Report*, October 3, 2011, http://www.hagstromreport.com/2011news_files/100311_lucas.html.

1. *Looking Ahead: Kansas and the 2012 Farm Bill*, Field Hearing before the Committee on Agriculture, Nutrition, and Forestry, US Senate, 112th Congress, 1st session, August 25, 2011.

2. David Mayhew, *Congress: The Electoral Connection* (New Haven, CT: Yale University Press, 1974), 49.

3. *Opportunities for Growth: Michigan and the 2012 Farm Bill*, Field Hearing before the Committee on Agriculture, Nutrition, and Forestry, US Senate, 112th Congress, 1st session, May 31, 2011.

4. Roberts quoted in *Looking Ahead: Kansas and the 2012 Farm Bill*, 6.

5. *Growing Forward: Michigan's Food and Agriculture Industry*, Michigan Department of Agriculture, 2015, https://www.michigan.gov/documents/mdard/1262-AgReport-2012_2_404589_7.pdf.

6. Kansas Department of Agriculture, https://agriculture.ks.gov/about-ksda/kansas-agriculture.

7. Roberts quoted in *Looking Ahead: Kansas and the 2012 Farm Bill*, 6.

8. Stabenow quoted in ibid., 5.

9. See, for example, Federalist 49–52, in James Madison, Alexander Hamilton, and John Jay, *The Federalist Papers*, ed. Clinton Rossiter (New York: New American Library, 1961).

10. See Charles O. Jones, *Introduction to the Study of Public Policy*, 2nd ed. (North Scituate, MA: Duxbury Press, 1977), 26.

11. John Kingdon, *Agendas, Alternatives, and Public Policy* (Boston: Little, Brown, 1984).

12. See Lawrence Jacobs and Theda Skocpol, *Health Care Reform and American Politics: What Everyone Needs to Know* (New York: Oxford University Press, 2010).

13. Kevin Donnelly and David Rochefort, "The Lessons of Lesson Drawing: How the Obama Administration Attempted to Learn from Failure of the Clinton Health Plan," *Journal of Policy History* 24, 2 (2012): 184–223.

14. See Paul Pierson, "Increasing Returns, Path Dependence, and the Study of Politics," *American Political Science Review* 94, 2 (2000): 251–267.

15. A notable exception is the federal transportation bill, whose provisions also lapse unless formally reauthorized. See Costas Panagopoulos and Joshua Shank, *All Roads Lead to Congress: The $300 Billion Fight over Highway Funding* (Washington, DC: CQ Press, 2008). See also E. Scott Adler and John Wilkerson, *Congress and the Politics of Problem Solving* (New York: Cambridge University Press, 2012).

16. On commodity programs, see Parke Wilde, *Food Policy in the United States: An Introduction* (New York: Routledge, 2011), chap. 2.

17. Earl O. Heady, *A Primer on Food, Agriculture, and Public Policy* (New York: Random House, 1967), 58.

18. Ibid., 60.

19. Christopher Bosso, *Pesticides and Politics: Life Cycle of a Public Issue* (Pittsburgh: University of Pittsburgh Press, 1987).

20. Devan McGranahan et al., "A Historical Primer on the US Farm Bill: Supply Management and Conservation Policy," *Journal of Soil and Water Conservation* 68, 3 (2013): 67–73A.

21. Jerry Hagstrom, "Farm Bill's Roots in Old Laws Should Be Sustained," *National Journal*, July 21, 2013, http://www.nationaljournal.com/columns/farm-bills-roots-in-old-laws-should-be-sustained-20130721.

22. John Schnittker, "The 1972–1973 Food Price Spiral," *Brookings Papers on Economic Activity* 2 (1973): 498–507.

23. See Michael Pollan, *The Omnivore's Dilemma: A Natural History of Four Meals* (New York: Penguin, 2006), 51–52.

24. Tom Philpott, "The Short-term Solution that Stuck: Where Farm Subsidies Came from and Why They're Still Here," *Grist*, January 30, 2007, http://www.grist.org/comments/food/2007/01/30/farm_bill2/.

25. Nadine Lehrer, "From Competition to National Security: Policy Change and Policy Stability in the 2008 Farm Bill" (Diss., University of Minnesota, 2008), 10.

26. For a short official history of the food stamp program, see Food and Nutrition Service, USDA, http://www.fns.usda.gov/snap/short-history-snap.

27. Christopher Bosso and Nicole Tichenor, "Eating and the Environment: Ecological Implications of Food Production," in *Environmental Policy: New Directions for the 21st Century*, 9th ed., ed. Norman Vig and Michael Kraft (Washington, DC: CQ Press, 2015), 194–214.

28. Critics include Pollan, *Omnivore's Dilemma*; Eric Schlosser, *Fast Food Nation: The Dark Side of the All-American Meal* (New York: HarperCollins, 2002); Marion Nestle, *Food Politics: How the Food Industry Influences Nutrition and Health* (Berkeley: University of California Press, 2002).

29. William P. Browne, introduction to *The New Politics of Food*, ed. Don F. Hadwinger and William P. Browne (Lexington, MA: D. C. Heath, 1978), 3.

30. See Jeffrey M. Berry, *Feeding Hungry People: Rulemaking in the Food Stamp Program* (New Brunswick, NJ: Rutgers University Press, 1984); Joe Richardson, *A*

Concise History of the Food Stamps Program, Congressional Research Service Report No. 79-244, 1979, 3–4.

31. See Lyle Scherz and Otto Doerring, *The Making of the 1996 Farm Act* (Ames: Iowa State University Press, 1999).

32. Fred Kirschenmann, "A Food and Farm Bill for the 21st Century," in *Food Fight: A Citizen's Guide to the Next Food and Farm Bill*, ed. Daniel Imhoff (Healdsburg, CA: Watershed Media, 2012), p. 9.

33. Lehrer, "From Competition to National Security," 40–41.

34. Scherz and Doerring, *Making of the 1996 Farm Act*, 118.

35. Lehrer, "From Competition to National Security," 40–41.

36. Goeffrey Becker and Jasper Womach, *The 2002 Farm Bill: Overview and Status*, Congressional Research Service Report No. RL31195, September 3, 2002, 29.

37. Christopher McGrory Klyza and David Sousa, *American Environmental Policy, 1990–2006: Beyond Gridlock* (Cambridge, MA: MIT Press, 2008). In *Congress and the Politics of Problem Solving*, Adler and Wilkerson argue that Congress has increased the use of "compulsory" statutes precisely to force itself to revisit policy areas over time.

38. Hagstrom, "Farm Bill's Roots in Old Laws Should Be Sustained."

39. Timothy A. Huelskamp, "Congressional Change: Committees on Agriculture in the U.S. Congress" (Ph.D. diss., American University, 1995), 8.

CHAPTER FOUR. WHATEVER IT TAKES: FARMERS, FOOD STAMPS, AND COALITIONS OF CONVENIENCE

Epigraph: Marion Nestle, interview, in Louisa Kasdon, "5 Course with Marion Nestle," *Stuff Boston*, January 16, 2012, http://stuffboston.com/2012/01/16/5-courses-with-marion-nestle#.Vb-dsHjiBa8.

1. See http://agriculture.house.gov/farmbill.

2. *Future of U.S. Farm Policy: Formulation of the 2012 Farm Bill*, Hearings before the Committee on Agriculture, US House of Representatives, 112th Congress, 2nd session, serial 112-30, 2012.

3. "Let's Talk about the Farm Bill," *Museum of Science Magazine*, Spring 2012, 6, http://www.mos.org/sites/dev-elvis.mos.org/files/docs/advancement/mos_magazine_spring-2012.pdf.

4. See Marion Nestle, *Soda Politics: Taking on Big Soda (and Winning)* (New York: Oxford University Press, 2015); David Bell, Stacy Bell, and George Blackburn, *How Government Shapes the American Diet*, Harvard Business School Report 9-504-064, March 2004.

5. K. L. Robinson, *The 1985 Farm Bill: The Political Process and Some Possible Outcomes*, Cornell Agricultural Economics Staff Paper 85-8, April 1985, 1.

6. Wesley McCune, *The Farm Bloc* (New York: Greenwood Press, 1968), 262, emphasis added.

7. David Danbom, *The Resisted Revolution: Urban America and the Industrialization of Agriculture, 1900–1930* (Ames: Iowa State University Press, 1979).

8. Christiana Campbell, *The Farm Bureau and the New Deal* (Urbana: University of Illinois Press, 1962).

9. Graham K. Wilson, *Interest Groups in America* (New York: Oxford University Press, 1981), 20.

10. Danbom, *Resisted Revolution*, 189.

11. McCune, *Farm Bloc*, 132.

12. John Mark Hansen, *Gaining Access: Congress and the Farm Lobby, 1919–1981* (Chicago: University of Chicago Press, 1991), chap. 2.

13. Robert Salisbury, "An Exchange Theory of Interest Groups," *Midwest Journal of Political Science* 13 (1969): 1–32.

14. E. E. Schattschneider, *The Semi-Sovereign People: A Realist's View of Democracy in America* (Hinsdale, IL: Dryden Press, 1960).

15. See James Q. Wilson, *Political Organizations* (New York: Basic Books, 1973), chap. 16.

16. Hansen, *Gaining* Access, 107–110.

17. Jeff Benedict, *Poisoned: The True Story of the Deadly E. Coli Outbreak that Changed the Way Americans Eat* (Buena Vista, VA: Inspire Books, 2011).

18. William Baumol, *Welfare Economics and the Theory of the State* (Cambridge, MA: Harvard University Press, 1952).

19. Gary Williams and Oral Capps Jr., "Overview: Commodity Checkoff Programs," *Choices* 21, 2 (2006): 53–54.

20. Mancur Olson, *The Logic of Collective Action* (Cambridge, MA: Harvard University Press, 1964).

21. See Charles O. Jones, "Representation in Congress: The Case of the House Agriculture Committee," *American Political Science Review* 55 (June 1961): 358–367.

22. Hansen, *Gaining Access*, 175.

23. Douglas Cater, *Power in Washington: A Critical Look at Today's Struggle to Govern in the Nation's Capital* (New York: Random House, 1964), 159.

24. J. Leiper Freeman, *The Political Process: Executive Bureau–Legislative Committee Relations*, rev. ed. (New York: Random House, 1966), 11.

25. Cater, *Power in Washington*, 158, emphasis added.

26. George Godwin Jr., *The Little Legislatures: The Committees of Congress* (Amherst: University of Massachusetts Press, 1970).

27. Hansen, *Gaining Access*, 171. See also William P. Browne et al., *Sacred Cows and Hot Potatoes: Agrarian Myths in Agricultural Policy* (Boulder, CO: Westview Press, 1992).

28. See William P. Browne, *Private Interests, Public Policy, and American Agriculture* (Lawrence: University Press of Kansas, 1988); Parke Wilde, *Food Policy in the United States: An Introduction* (New York: Routledge, 2013); Bruce Gardner, *Ameri-*

can Agriculture in the Twentieth Century: How It Flourished and What It Cost (Cambridge, MA: Harvard University Press, 2002).

29. Hansen, *Gaining Access*, 172.

30. John Ferejohn, "Logrolling in an Institutional Context: A Case Study of Food Stamp Legislation," in *Congress and Policy Change*, ed. Gerald C. Wright Jr., Leroy N. Rieselbach, and Lawrence C. Dodd (New York: Agathon Press, 1986), 223–253.

31. Randall B. Ripley, "Legislative Bargaining and the Food Stamp Act, 1964," in *Congress and Urban Problems: A Casebook on the Legislative Process*, ed. Frederic C. Cleaveland (Washington, DC: Brookings Institution, 1969), 293.

32. Timothy A. Huelskamp, "Congressional Change: Committees on Agriculture in the U.S. Congress" (Ph.D. diss., American University, 1995), 164.

33. Hansen, *Gaining Access*, 210.

34. Huelskamp, "Congressional Change," 192.

35. Nestle interview, in Kasdon, "5 Course with Marion Nestle."

36. John G. Peters, "The 1977 Farm Bill: Coalitions in Congress," in *The New Politics of Food*, ed. Don F. Hadwiger and William P. Browne (Lexington, MA: Lexington Books, 1978), 25.

37. For example, see Theodore J. Lowi, "Four Systems of Policy, Politics, and Choice," *Public Administration Review* 32, 298 (1972): 299–300; Randall B. Ripley and Grace A. Franklin, *Congress, the Bureaucracy, and Public Policy*, 2nd ed. (Homewood, IL: Dorsey Press, 1979); William P. Browne, *Cultivating Congress: Constituents, Issues, and Interests in Agricultural Policymaking* (Lawrence: University Press of Kansas, 1995).

38. Huelskamp, "Congressional Change," 95.

39. Ibid., 260–261.

40. Ibid., 317.

41. Ibid., 355.

42. Jerry Hagstrom, "Farm Team Goes to Bat for Food Stamps," *National Journal*, March 11, 1995, 623.

43. Jerry Hagstrom, "Proposal to Split Farm Bill Divides Congress," *National Journal Daily*, July 8, 2013, http://www.nationaljournal.com/daily/proposal-to-split-farm-bill-divides-congress-20130707?print=true.

44. Paul A. Sabatier and Christopher M. Weible, "The Advocacy Coalition Framework," in *Theories of the Policy Process*, 2nd ed., ed. Paul A. Sabatier (Boulder, CO: Westview Press, 2007), 196.

45. Although the USDA refuses to publish data on specific retailers, the greatest proportion of food stamps is redeemed at supermarkets and big-box stores like Walmart. See Laura Castner and Juliette Kenke, *Benefit Redemption Patterns in the Supplemental Nutrition Assistance Program*, Food and Nutrition Service, USDA, February 2011. See also Krissy Clark, "The Secret Life of a Food Stamp," *Slate*, April 14, 2014, http://www.slate.com/articles/business/moneybox/2014/04/big_box_stores_make_billions_off_food_stamps_often_it_s_their_own_workers.html.

46. Michael Pollan, "The Food Movement, Rising," *New York Times*, June 10, 2010, http://www.nybooks.com/articles/2010/06/10/food-movement-rising/.

47. Bruce Yandle, "Bootleggers and Baptists: The Education of a Regulatory Economist," *Regulation*, May–June 1983, 12–16.

CHAPTER FIVE. "STOP THE SPENDING": BUDGET POLITICS AND THE "SECRET" FARM BILL

Epigraph: Quoted in Tim Fernholz, "GOP Freshmen: 'Full Steam Ahead,'" *National Journal*, January 21, 2011, Factiva, Document NTLJ000020110116e71d0000g.

1. Michael Barone and Chuck McCutcheon, *The Almanac of American Politics, 2014* (Chicago: National Journal/University of Chicago Press, 2013), 2–3.

2. For an assessment, see Thomas E. Mann and Norman J. Ornstein, *It's Even Worse than It Looks: How the American Constitutional System Collided with the New Politics of Extremism* (New York: Basic Books, 2012).

3. Jerry Hagstrom, "Breaking Ground," *National Journal*, January 17, 2011, Factiva, Document CNGA000020110118e71h00006. More generally, see Thomas Frank, *What's the Matter with Kansas? How Conservatives Won the Heart of America* (New York: Henry Holt, 2004).

4. Christopher Ingraham, *Historical House Ideology and Party Unity, 35th–113th Congress (1857–2014)*, Brookings Institution, http://www.brookings.edu/research/interactives/2013/historical-house-ideology-and-party-unity#.

5. Fernholz, "GOP Freshmen: 'Full Steam Ahead.'"

6. Norman Ornstein, "Four Really Dumb Ideas that Should Be Avoided," *Roll Call*, January 26, 2011, http://www.rollcall.com/issues/56_72/-202782-1.html.

7. Jill Lawrence, *Profiles in Negotiation: The 2014 Farm and Food Stamp Deal*, Center for Effective Public Management Paper (Washington, DC: Brookings Institution, 2015), 4.

8. This is nicely depicted in Costas Panagopoulos and Joshua Shank, *All Roads Lead to Congress: The $300 Billion Fight over Highway Funding* (Washington, DC: CQ Press, 2008).

9. See the Open Secrets database operated by the Center for Responsive Politics and Taxpayers for Common Sense: https://www.opensecrets.org/earmarks/.

10. Scott Frisch and Sean Kelly, *Cheese Factories on the Moon: Why Earmarks Are Good for American Democracy* (New York: Routledge, 2011); Diana Evans, *Greasing the Wheels: Using Pork Barrel Projects to Build Majority Coalitions in Congress* (New York: Cambridge University Press, 2004).

11. See http://clerk.house.gov/evs/2002/roll123.xml; http://clerk.house.gov/evs/2008/roll315.xml.

12. Jennifer Steinhauer, "Congress to Return with G.O.P. Vowing to Alter Rules," *New York Times*, January 5, 2011, A14.

13. Jerry Hagstrom, "Stabenow: Senate Will Proceed on Its Own Schedule to

Write, Finish Farm Bill," *Hagstrom Report*, March 28, 2011, http://www.hagstromreport.com/2011news_files/032811_stabenow.html.

14. John O'Sullivan, "John Boehner Calls Debt Cap Increase Inevitable, McConnell Mum on Shutdown," *National Journal*, January 30, 2011, Factiva, Document NTLJ000020110131e71u00001.

15. David Herszenhorn, "Stopgap Bill Hamstrings Government Programs," *New York Times*, December 22, 2010, A19.

16. David Herszenhorn, "House Spending Impasse Raises Risk of Shutdown," *New York Times*, February 18, 2011, A16.

17. Editorial, "Here's an Easy One," *New York Times*, January 16, 2011, WK9.

18. P. J. Huffstetter, "Obama's Budget Would Cut Deeply into Farm Subsidies," *Los Angeles Times*, February 14, 2011.

19. "Here's an Easy One," WK9.

20. Jennifer Steinhauer, "Farm Subsidies Become Target Amid Spending Cuts," *New York Times*, May 7, 2011, A13.

21. Mark Bittman, "Don't End Agricultural Subsidies. Fix Them," *New York Times*, February 3, 2011, http://opinionator.blogs.nytimes.com/2011/03/01/dont-end-agricultural-subsidies-fix-them/.

22. "Here's an Easy One," WK9; Randy Krehbiel, "Rep. Frank Lucas' Farm Experiences Reflected in Farm Bill Proposals," *Tulsa World*, October 28, 2013.

23. Tim Fernholz, "Ag Committee Supports Cuts to Food Assistance, Not Farm Subsidies," *National Journal*, March 21, 2011.

24. Steinhauer, "Farm Subsidies Become Target," A13.

25. Ron Nixon, "In Battle over Subsidies, Some Farmers Say No," *New York Times*, June 23, 2011, A14.

26. Steinhauer, "Farm Subsidies Become Target," A13.

27. Nixon, "In Battle over Subsidies," A14.

28. Hagstrom, "Stabenow: Senate Will Proceed."

29. Paul Kane, Philip Rucker, and David Fahrenholdt, "Congress Agrees to Eleventh-Hour Budget Deal to Avert Government Shutdown," *Washington Post*, April 4, 2011, http://www.washingtonpost.com/politics/reid-says-impasse-based-on-abortion-funding-boehner-denies-it/2011/04/08/AFO40U1C_story.html.

30. Jackie Calmes, "Geithner Urges Congress to Increase Debt Limit," *New York Times*, January 6, 2011, A1; Jennifer Steinhauer, "Cantor Says House Won't Raise Debt Limit 'Without Serious Cuts,'" *New York Times*, January 18, 2011, A1.

31. Editorial, "The Super-committee Collapses," *New York Times*, November 22, 2011, A28.

32. Lori Montgomery and Paul Kane, "Debt-Limit Vote Is Canceled in House as Boehner, GOP Leaders Struggle to Gain Votes," *Washington Post*, July 28, 2011, http://www.washingtonpost.com/business/economy/boehner-other-gop-leaders-put-pressure-on-republicans-to-pass-debt-plan/2011/07/28/gIQARD5veI_story.html.

33. Brian Knowlton, "Last Three Democrats Named to Debt Committee," *New York Times*, August 11, 2011.

34. For a summary of the 2011 "debt crisis," see Mann and Ornstein, *It's Even Worse than It Looks*, 3–30.

35. Jerry Hagstrom, "Unusual Seeds for a Farm Bill," *National Journal*, October 24, 2011, Factiva, Document CNGA000020111026e7a00000e.

36. Jerry Hagstrom, "Subsidies Drive Farm Debate," *National Journal Daily*, May 14, 2012, Factiva, Document CNGA000020120516e85e0000c.

37. Jerry Hagstrom, "A Super Farm Bill?" *National Journal*, October 3, 2011, Factiva, Document CNGA000020111005e7a30000e.

38. Hagstrom, "Unusual Seeds for a Farm Bill."

39. Mark Bittman, "The Secret Farm Bill," *New York Times*, November 8, 2011, http://opinionator.blogs.nytimes.com/2011/11/08/the-secret-farm-bill/.

40. Shane Goldmacher and Dan Friedman, "House Approves $915 Billion Spending Plan," *National Journal*, December 16, 2011.

41. Hagstrom, "Unusual Seeds for a Farm Bill."

42. William Neuman, "Farmers Facing Loss of Subsidy May Get New One," *New York Times*, October 18, 2011, A1.

43. Comments by Rep. Frank Lucas, *The Future of U.S. Farm Policy: Formulation of the 2012 Farm Bill*, Hearings before the Committee on Agriculture, US House of Representatives, 112th Congress, 2nd session, March 9, 2012, serial no. 112-30, pt. 1, p. 24.

CHAPTER SIX. BUILDING A PATHWAY TO SIXTY:
THE SENATE MOVES FIRST

Epigraph: Quoted in Jerry Hagstrom, "Senate Votes to Proceed on Farm Bill," *Agweek*, June 7, 2012.

1. Jerry Hagstrom, "The Fate of the Farm Bill," *National Journal*, January 16, 2012, Factiva, Document CNGA000020120118e81g0000c.

2. Jerry Hagstrom, "Budget Priorities," *Agweek*, March 12, 2012.

3. Jerry Hagstrom, "Nutrition Programs Debated," *Agweek*, April 16, 2012.

4. Hagstrom, "Fate of the Farm Bill."

5. Jerry Hagstrom, "Ag Leaders Want Farm Bill This Year," *Agweek*, April 23, 2012.

6. Jerry Hagstrom, "Hope on the Farm," *National Journal*, March 5, 2012.

7. Charles Tilly, *Big Structures, Large Processes, Huge Comparisons* (New York: Russell Sage, 1984), 14, emphasis in original.

8. On the various legislators, see Michael Barone and Chuck McCutcheon, *The Almanac of American Politics 2014* (Chicago: University of Chicago Press, 2013).

9. *Hearings of the Committee on Agriculture, Nutrition, and Forestry*, 112th Congress, 1st session, May 26, 2011, 4.

10. Editorial, "Chipping Away at Gridlock," *New York Times*, October 11, 2011, 26.

11. Timothy A. Huelskamp, "Congressional Change: Committees on Agriculture in the U.S. Congress" (Ph.D. diss., American University, 1995), 108.

12. Jerry Hagstrom, "Commodity Title Taking Shape," *Agweek*, March 12, 2012.

13. William Neuman, "Farmers Facing Loss of Subsidy May Get New One," *New York Times*, October 18, 2011, A1.

14. Jerry Hagstrom, "Farm Groups Testify at Ag Hearing," *Agweek*, March 19, 2012; Hagstrom, "Ag Leaders Want Farm Bill This Year."

15. Jerry Hagstrom, "Conrad's Farm Bill Problems," *Agweek*, March 19, 2012.

16. Josh Voorhees, "Heat or Eat," *Slate*, July 28, 2014, http://www.slate.com/articles/news_and_politics/politics/2014/07/heat_and_eat_republicans_keep_trying_to_close_a_snap_loophole_that_doesn.html.

17. Jerry Hagstrom, "Ryan Budget Cuts $180 Billion from Farm Bill Programs," *Agweek*, March 22, 2012.

18. Hagstrom, "Ag Leaders Want Farm Bill This Year."

19. Jerry Hagstrom, "Dairy Wars," *National Journal*, March 14, 2011.

20. Jerry Hagstrom, "Harvest Time," *National Journal*, April 23, 2012.

21. Tina Susman, "Lottery Winner Who Drew Outrage for Getting Welfare Is Found Dead," *Los Angeles Times*, October 2, 2012.

22. For a transcript, see http://www.ag.senate.gov/hearings/business-meeting-farm-bill-markup.

23. David Bennett, "Criticism Heavy during Farm Bill Markup," *Delta Farm Press*, April 27, 2012, Factiva, Document DLFP000020120428e84r00002.

24. Jerry Hagstrom, "Senate Ag Committee Passes Farm Bill," *Agweek*, April 30, 2012. See also Alexandria Jaffe, "Mitch McConnell's Agriculture Absences Raise Questions," *Hill*, August 19, 2014, http://thehill.com/blogs/ballot-box/senate-races/215545-mcconnells-agriculture-absences-raise-eyebrows.

25. See Keith Good for FarmPolicy.com, http://farmpolicy.com/2012/04/27/farm-bill-appropriations-and-regulations/#more-7203.

26. Jerry Hagstrom, "Farm Bill Field Hearing Addresses Dairy, Specialty Crops," *Agweek*, March 12, 2012.

27. Jerry Hagstrom, "Subsidies Drive Farm Debate," *National Journal Daily*, May 14, 2012, Factiva, Document CNGA000020120516e85e0000c.

28. Jerry Hagstrom, "Standing by Agriculture, *Agweek*, March 12, 2012.

29. Hagstrom, "Harvest Time."

30. As this applied to immigration reform legislation, see Molly K. Hooper, "Boehner: I'm Not for a Comprehensive Solution," *Hill*, June 19, 2013, 1.

31. Actually, there were fifty-one Democrats plus Bernie Sanders of Vermont, then an Independent.

32. For one view, see Thomas E. Mann and Norman J. Ornstein, *It's Even Worse than It Looks: How the American Constitutional System Collided with the New Politics of Extremism* (New York: Basic Books, 2012).

33. Jack Torrey, "A Broken Congress—The Senate: GOP Filibusters; Dems 'Fill the Tree,'" *Columbus Post Dispatch*, June 1, 2014, http://www.dispatch.com/content/stories/local/2014/06/01/gop-filibusters-dems-fill-the-tree.html. The definitive source is

Floyd M. Riddick and Alan S. Frumin, *Riddick's Senate Procedure: Precedents and Practices*, 101st Congress, 1st session, S. Doc. 101-28 (Washington, DC: Government Printing Office, 1992), 74–89, http://www.gpo.gov/fdsys/pkg/GPO-RIDDICK-1992/pdf/GPO-RIDDICK-1992-7.pdf.

34. Meredith Shiner, "Broad Coalition of Senators Pushes Farm Bill," *Roll Call*, May 15, 2012, http://www.rollcall.com/news/broad_coalition_of_senators_pushes_farm_bill-214517-1.html.

35. Jerry Hagstrom, "Saving the Farm Bill," *National Journal*, May 12, 2012, Factiva, Document CNGA000020120523e85l00007.

36. Meredith Shiner, "Floor Fight Will Test Farm Bill," *Roll Call*, June 5, 2012, http://www.rollcall.com/issues/57_145/Floor-Fight-Will-Test-Farm-Bill-215041-1.html.

37. Meredith Shiner, "John McCain Throws 'Sequester' Wrench into Farm Bill," *Roll Call*, June 6, 2012, http://www.rollcall.com/news/john_mccain_throws_sequester_wrench_into_farm_bill-215124-1.html.

38. Ron Nixon, "Senate Advances Farm Bill," *New York Times*, June 7, 2012, http://nyti.ms/NOeLiv; Hagstrom, "Senate Votes to Proceed on Farm Bill."

39. Meredith Shiner, "Farm Bill Faces Uphill Battle," *Roll Call*, June 12, 2012; Jacqui Fatka, "Non-germane Amendments Hold Farm Bill Hostage," *Farm Futures*, June 15, 2012, http://farmfutures.com/blogs-non-germane-amendments-hold-farm-bill-hostage-3341.

40. Jerry Hagstrom, "Senate Rejects Amendment to End Sugar Program," *Agweek*, June 14, 2012.

41. Jerry Hagstrom, "Farm Bill Stalled," *Agweek*, June 15, 2012.

42. Meredith Shiner, "Senate Agrees on Way Forward on Farm Bill," *Roll Call*, June 18, 2012, http://www.rollcall.com/news/Senate-Agrees-on-Way-Forward-on-Farm-Bill-215476-1.html; Ron Nixon, "Stack of Farm Proposals Is Coming up for Votes," *New York Times*, June 20, 2012, A19, http://nyti.ms/N4U5k2.

43. Jerry Hagstrom, "Deal Makers," *National Journal*, June 18, 2012.

44. Jerry Hagstrom, "Stabenow Lessons," *National Journal*, June 25, 2012.

45. Jerry Hagstrom, "Senate Passes Farm Bill," *Agweek*, June 21, 2012.

46. Roll call on S. 3420, http://www.senate.gov/legislative/LIS/roll_call_lists/roll_call_vote_cfm.cfm?congress=112&session=2&vote=00164.

47. Ron Nixon, "Senate Passes Farm Bill with Bipartisan Support," *New York Times*, June 22, 2012, A16.

48. Hagstrom, "Senate Passes Farm Bill."

CHAPTER SEVEN. WE'RE ON THE ROAD TO NOWHERE

For my choice of chapter title, apologies to David Byrne. Epigraph: Quoted in Timothy A. Huelskamp, "Congressional Change: Committees on Agriculture in the U.S. Congress" (Ph.D. diss., American University, 1995), 108.

1. Meredith Shiner, "Senate's Farm Bill Journey Ends in Passage," *Roll Call*, June 21, 2012.

2. Ron Nixon, "Split among House Republicans over How Deeply to Cut May Delay Farm Bill," *New York Times*, July 13, 2012, A13.

3. Daniel Newhauser, "Farm Bill Sparks Infighting in Parties," *Roll Call*, July 11, 2012.

4. Jerry Hagstrom, "House Ag Committee Releases Its Farm Bill," *Agweek*, July 9, 2012.

5. Newhauser, "Farm Bill Sparks Infighting."

6. Nixon, "Split among House Republicans."

7. Proposed amendments can be viewed at https://archives-agriculture.house.gov/markup/consider-2012-farm-bill.

8. Daniel Newhauser, "Members Prod House Leaders for Vote on Stalled Farm Bill," *Roll Call*, July 20, 2012.

9. *Federal Agriculture Reform and Risk Management Act of 2012, Report of the Committee on Agriculture Together with Additional and Dissenting Views* [to Accompany H.R. 6083], 112th Congress, 2nd session, Rept. 112-669, September 13, 2012, 679, http://www.gpo.gov/fdsys/pkg/CRPT-112hrpt669/pdf/CRPT-112hrpt669.pdf.

10. Nixon, "Split among House Republicans."

11. The Madison Project, http://www.conservativevotingrecords.com/agriculture-committee-voting-report-on-farm-bill/.

12. Jerry Hagstrom, "House Considers One-Year Farm Bill Extension," *Agweek*, July 27, 2012.

13. Ibid.

14. Jerry Hagstrom, "House Ag Committee Approves Farm Bill," *Agweek*, July 12, 2012, Factiva, Document KRTAW00020120717e87c00009.

15. Jennifer Steinhauer, "Enduring Drought, Farmers Draw the Line at Congress," *New York Times*, August 13, 2012, http://www.nytimes.com/2012/08/13/us/politics/drought-driven-voters-vent-anger-over-farm-bill.html.

16. Jerry Hagstrom, "House Urged to Act on Farm Bill," *Agweek*, July 20, 2012.

17. Newhauser, "Members Prod House Leaders."

18. Hagstrom, "House Considers One-Year Farm Bill Extension."

19. Jerry Hagstrom, "House Passes Disaster Assistance," *Agweek*, August 2, 2012.

20. Steinhauer, "Enduring Drought," A7; Humberto Sanchez, "Campaign Fodder in Stalled Farm Bill," *Roll Call*, September 19, 2012.

21. Humberto Sanchez, "Kristi Noem Not Happy about Farm Bill's Fate," *Roll Call*, September 12, 2012, emphasis added.

22. Erik Wasson, "House Leaders Try to Move Three-Month Extension of Farm Bill," *Hill*, September 14, 2012.

23. Humberto Sanchez, "Democrats Weigh Farm Bill Threat; Senators Consider Letting Agriculture Programs Expire to Spur House Action," *Roll Call*, September 12, 2012.

24. David Roger, "The Kick-the-Can Congress," *Politico*, July 23, 2012, http://www.politico.com/story/2012/07/the-kick-the-can-congress-078832.

25. Brad Plumer, "Congress Just Let the Farm Bill Expire. It's Not the End of the World . . . Yet," Wonk Blog, *Washington Post*, October 1, 2012.

26. Hagstrom, "House Passes Disaster Assistance."

27. Jerry Hagstrom, "Vilsack Urges Strategy in Political Fights," *Agweek*, December 17, 2012.

28. The winner in Maine was Independent Angus King, who, like Vermont's Bernie Sanders, would caucus with the Democrats.

29. Sanchez, "Campaign Fodder in Stalled Farm Bill."

30. Larry Dreiling, "Huelskamp Removed from House Ag Committee," *High Plains Journal*, December 10, 2012.

31. Ibid.

32. Dan Voorhis, "Boehner Ousts Huelskamp from Agriculture and Budget Committees," *Wichita Eagle*, December 4, 2012.

33. Dreiling, "Huelskamp Removed."

34. Plumer, "Congress Just Let the Farm Bill Expire."

35. Hagstrom, "Vilsack Urges Strategy."

36. Jerry Hagstrom, "Farm Bill's Fate Rests on Cliff Negotiations," *National Journal Daily*, December 3, 2012.

37. Renee Schoof, "Food Stamp Dispute Threatens Final Push for Farm Bill," *Wichita Eagle*, December 5, 2012.

38. Jerry Hagstrom, "Farm Bill Gridlock," *Agweek*, December 17, 2012.

39. Jerry Hagstrom, "Farm Bill, Dairy Program Prospects Remain Uncertain," *Agweek*, December 31, 2012.

40. Ibid.

41. David Rogers, "30-Day Farm Bill Called 'Poor Joke,'" *Politico*, December 30, 2012.

42. David Rogers, "Fiscal Cliff Deal Includes Farm Bill Extension," *Politico*, January 1, 2013; Ron Nixon, "Tax Bill Passed by Senate Includes Farm Bill Extension," *New York Times*, January 1, 2013, http://www.nytimes.com/interactive/us/politics/debt-reckoning.html#sha=48e66885d.

43. Nixon, "Tax Bill Passed by Senate."

44. David Henry, "Peterson Lets Fly: Obama, Leaders, Don't Care about Farm Bill," *Minnesota Post*, January 1, 2013, https://www.minnpost.com/dc-dispatches/2013/01/peterson-lets-fly-obama-leaders-don-t-care-about-farm-bill; Rogers, "Fiscal Cliff Deal."

45. Roll call 659, January 1, 2013, http://clerk.house.gov/evs/2012/roll659.xml.

CHAPTER EIGHT. SNAP

Epigraph: John A. Schnittker, "Farm Payments," *New Republic*, June 27, 1970, 11, as quoted in John Mark Hansen, *Gaining Access: Congress and the Farm Lobby, 1919–1981* (Chicago: University of Chicago Press, 1991), 173.

1. Ben Terris, "There May Be No Way to Silence Republican Outcast Tim Huelskamp," *National Journal*, January 23, 2013, 5.

2. For the sake of brevity, any reference to "Senate Democrats" includes King and Sanders, who caucused with and generally voted as Democrats.

3. Dan Voorhis, "Pat Roberts Bumped from Leadership Post on Senate Ag Committee," *Wichita Eagle*, January 4, 2013, http://www.kansas.com/news/politics-government/article1105849.html; Jerry Hagstrom, "New Congress Faces Old Problems in Effort to Pass Long-Term Farm Bill," *National Journal*, January 21, 2013.

4. On relative committee prestige, see Charles Stewart III, "The Value of Committee Assignments in Congress since 1994," MIT Political Science Department Research Paper 2012-7, Midwest Political Science Association, available at http://ssrn.com/abstract=2035632.

5. Timothy A. Huelskamp, "Congressional Change: Committees on Agriculture in the U.S. Congress" (Ph.D. diss., American University, 1995), chap. 5.

6. *National Journal* vote ratings are available at http://www.nationaljournal.com/2013-vote-ratings.

7. SNAP data obtained from the American Community Survey, 2009–2011, http://www.fns.usda.gov/Ora/SNAPCharacteristics/default.htm.

8. Michael Barone and Chuck McCutcheon, *The Almanac of American Politics 2014* (Chicago: University of Chicago Press, 2013), 820.

9. Ron Nixon, "Farm Subsidy Recipient Backs Food Stamp Cuts," *New York Times*, May 23, 2013, A20.

10. David Rogers, "Peterson Pushes Boehner for Farm Bill Vote," *Politico*, January 4, 2013. The original letter is at http://democrats.agriculture.house.gov/inside/Pubs/01-03-2012%20Boehner%20Farm%20Bill.pdf.

11. David Rogers, "New Farm Bill Leans on Food Stamps," *Politico*, May 6, 2013.

12. Jonathan Weisman, "In Reversal, House G.O.P. Agrees to Lift Debt Limit," *New York Times*, January 19, 2013, A1.

13. David Rogers, "Agriculture Has Slipped from D.C.'s Radar Screen," *Politico*, February 25, 2013.

14. Ron Nixon, "Record Taxpayer Cost Is Seen for Crop Insurance," *New York Times*, January 16, 2013, A16.

15. David Rogers, "New Estimates Cut Farm Bill Savings," *Politico*, March 1, 2013.

16. Jerry Hagstrom, "Food Stamps Are Key Component to Getting Farm Bill Passed," *National Journal*, April 10, 2013.

17. Jerry Hagstrom, "Congress Poised to Move on Farm Bill," *National Journal*, May 12, 2013.

18. David Rogers, "Draft Farm Bill Aids Pork, Beef Lobbies," *Politico*, May 9, 2013.

19. David Rogers, "House GOP Rolls out New Farm Bill," *Politico*, May 10, 2013.

20. Rogers, "New Farm Bill Leans on Food Stamps."

21. Ron Nixon, "Senate Panel Approves Farm Bill," *New York Times*, May 14, 2013.

22. David Rogers, "Farm Bill Clears Senate Agriculture Committee," *Politico*, May 14, 2013.

23. Committee on Agriculture, House of Representatives, *Federal Agriculture Reform and Risk Management Act of 2013*, 113th Congress, 1st session, House Report 113-92, pts. 1–3; Ron Nixon, "House Agriculture Committee Approves Farm Bill," *New York Times*, May 16, 2013.

24. David Rogers, "House Panel Approves Farm Bill," *Politico*, May 15, 2013.

25. Hagstrom, "Food Stamps Are Key Component."

26. David Rogers, "Senate Debates Farm Bill," *Politico*, May 20, 2013.

27. Ron Nixon, "As Obama Pushes Overhaul of Food Aid, Panels in Congress Favor Smaller Changes," *New York Times*, May 18, 2013, A13.

28. Ascribed to any number of people, including former senator Morris K. Udall (D-AZ).

29. Record vote 131, Senate amendment 931 to S. 954, May 22, 2013, http://www.senate.gov/legislative/LIS/roll_call_lists/roll_call_vote_cfm.cfm?congress=113&session=1&vote=00131.

30. David Rogers, "Farm Bill Beats Back Foes," *Politico*, May 21, 2013; record vote 130, Senate amendment 948 to S. 954, May 22, 2013, http://www.senate.gov/legislative/LIS/roll_call_lists/roll_call_vote_cfm.cfm?congress=113&session=1&vote=00130.

31. Record vote 132, Senate amendment 960 to S. 954, May 22, 2013, http://www.senate.gov/legislative/LIS/roll_call_lists/roll_call_vote_cfm.cfm?congress=113&session=1&vote=00132.

32. David Rogers, "Farm Bill: Sugar Interests Win Senate Vote," *Politico*, May 22, 2013.

33. David Rogers, "Hemp, GMO Labeling Debates Hit Farm Bill," *Politico*, May 23, 2013; record vote 135, Senate amendment 965 to S. 954, May 22, 2013, http://www.senate.gov/legislative/LIS/roll_call_lists/roll_call_vote_cfm.cfm?congress=113&session=1&vote=00135. Of note, in 2016 Congress passed and President Obama signed a law creating a federal genetically modified food–labeling program that preempted state labeling laws.

34. Record vote 139, Senate amendment 953 to S. 954, May 23, 2013, http://www.senate.gov/legislative/LIS/roll_call_lists/roll_call_vote_cfm.cfm?congress=113&session=1&vote=00139.

35. David Rogers, "Crunch Time for Senate Farm Bill," *Politico*, June 5, 2013.

36. David Rogers, "Farm Bill Advances in Senate," *Politico*, June 6, 2013; David Rogers, "Strong Win in Senate for Farm Bill Test Vote," *Politico*, June 6, 2013.

37. David Rogers, "Farm Bill Passes Senate with Bipartisan Majority," *Politico*, June 10, 2013.

38. Jerry Hagstrom, "Lawmakers Need to Show Strong Leadership to Move Forward with Farm Bill," *National Journal*, June 9, 2013; Ron Nixon, "Senate Passes Farm Bill; House Vote Is Less Sure," *New York Times*, June 11, 2013, A10, http://nyti.ms/19180Ex.

39. Ramsey Cox, "Senate Passes Farm Bill in 66–27 Vote," *Hill*, June 10, 2012, http://thehill.com/blogs/floor-action/senate/304585-senate-votes-66-27-to-pass-a-five-year-farm-bill.

40. Rogers, "Farm Bill Advances in Senate."

41. Derek Welbank and Alan Bjerga, "Senate Votes to Scale Back Farm Subsidies," *Bloomberg Business*, June 10, 2013, http://www.bloomberg.com/news/articles/2013-06-10/senate-votes-to-scale-back-farm-subsidies.

42. David Rogers, "Heritage Ads Attack Farm Bill," *Politico*, May 29, 2013.

43. Welbank and Bjerga, "Senate Votes to Scale Back Farm Subsidies."

44. Charles Abbott, "Senate Passes Farm Bill; Food Stamp Fight Looms in House," *Reuters*, June 10, 2013, http://www.reuters.com/article/us-usa-agriculture-farm-bill-idUSBRE95915T20130611.

45. David Rogers, "Tom Vilsack Urges House to Plow Ahead on Farm Bill," *Politico*, June 12, 2013.

46. David Rogers, "Nancy Pelosi 'Not Likely' to Back Farm Bill," *Politico*, June 18, 2013.

47. Jerry Hagstrom, "Farm Bill Could Pass the House Next Week, Agriculture Committee Chairman Says," *National Journal*, June 13, 2013.

48. Roll-call vote 256 on H.R. 1947, June 19, 2013, http://clerk.house.gov/evs/2013/roll256.xml.

49. David Rogers, "Farm Bill Advances in House," *Politico*, June 19, 2013.

50. House amendment 197 on H.R. 1947, June 19, 2013, https://www.congress.gov/amendment/113th-congress/house-amendment/197.

51. Jill Lawrence, *Profiles in Negotiation: The 2014 Farm and Food Stamp Deal* (Center for Effective Public Management, Brookings Institution, 2015), 10.

52. House amendment 47 on H.R. 1947, roll-call vote 276, June 19, 2013, http://clerk.house.gov/evs/2013/roll276.xml. See David Rogers, "House Set to Begin on Farm Bill," *Politico*, June 16, 2013; Ron Nixon, "Opposition to House Farm Bill Spans Political Spectrum," *New York Times*, June 17, 2013.

53. House amendment 100 on H.R. 1947, roll-call vote 282, June 20, 2013, http://clerk.house.gov/evs/2013/roll282.xml.

54. House amendment 90 on H.R. 1947, June 20, 2013, http://clerk.house.gov/evs/2013/roll278.xml.

55. House amendment 89 on H.R. 1947, June 20, 2013, http://clerk.house.gov/evs/2013/roll281.xml.

56. House amendment 101 on H.R. 1947, June 20, 2013, http://clerk.house.gov/evs/2013/roll283.xml.

57. Lawrence, *Profiles in Negotiation*, 10.

58. House amendment 102 on H.R. 1947, June 20, 2013, http://clerk.house.gov/evs/2013/roll284.xml. See also Jerry Hagstrom, "Farm Leaders Share Blame for House Debacle," *National Journal*, June 23, 2013.

59. Billy Chase, "House Farm Bill Suffers Stunning Defeat as Finger Pointing Be-

gins," *National Journal*, June 20, 2013; David Rogers, "How the Farm Bill Failed," *Politico*, June 23, 2013.

60. Michael McAuliffe, Arthur Delaney, and Sabrina Saddiqui, "Food Stamp Cuts Derail Farm Bill," *Huffington Post*, June 22, 2013.

61. Chase, "House Farm Bill Suffers Stunning Defeat"; Rogers, "How the Farm Bill Failed."

62. Editorial, "Farm Bill's Death Provokes Few Tears," *Washington Post*, June 21, 2013.

63. Ron Nixon, "House Defeat of Farm Bill Lays Bare Rift in G.O.P.," *New York Times*, June 21, 2013, A12, http://nyti.ms/19VCACq.

64. David Rogers, "House GOP Pulls Agriculture Spending Bill," *Politico*, June 24, 2013.

65. Jerry Hagstrom, "Proposal to Split Farm Bill Divides Congress," *National Journal Daily*, July 8, 2013, http://www.nationaljournal.com/daily/proposal-to-split-farm-bill-divides-congress-20130707?print=true.

66. Tal Kopan, "Obama Threatens Veto on Farm Bill," *Politico*, July 11, 2013, http://politi.co/15imXPN.

67. David Rogers, "Farm Bill 2013: House Narrowly Passes Pared-Back Version," *Politico*, July 11, 2013; roll-call vote 353, H.R. 2642, July 11, 2013, http://clerk.house.gov/evs/2013/roll353.xml.

68. Mary Claire Jalonek, "House OKs Scaled-Down Farm Bill Sans Food Stamps," *Huffington Post*, July 11, 2013, http://www.huffingtonpost.com/huff-wires/20130711/us-farm-bill/?utm_hp_ref=homepage&ir=homepage.

69. Collin Peterson, Floor Statement in Opposition to H.R. 2642, July 11, 2013, http://democrats.agriculture.house.gov/press/PRArticle.aspx?NewsID=1181.

70. Jerry Hagstrom, "House Passage of Stripped-Down Farm Bill Leaves Many Questions Unanswered," *National Journal*, July 11, 2013.

71. Arthur Delaney, "Farm Bill without Food Stamps Should Please Liberal: Frank Lucas," *Huffington Post*, July 11, 2013.

72. See https://www.congress.gov/congressional-record/2013/07/18/senate-section/article/S5794-1.

73. David Rogers, "House GOP Seeks Cuts in Food Stamps," *Politico*, September 16, 2013.

74. David Rogers, "House Approves Plan to Cut Food Stamps," *Politico*, September 19, 2013.

75. House Report 113-231, https://www.congress.gov/congressional-report/113th-congress/house-report/231; roll-call vote 493 on H. Res. 361, September 28, 2013, http://clerk.house.gov/evs/2013/roll493.xml.

CHAPTER NINE. IN CONFERENCE

Epigraph: Quoted in David Rogers, "House-Senate Farm Bill Talks OK'd," *Politico*, October 11, 2013.

1. Noah Bierman, "Boehner Pulled from Two Directions," *Boston Globe*, October 3, 2013.

2. Jonathon Weisman and Jeremy Peters, "Government Shuts Down in Budget Impasse," *New York Times*, October 1, 2013, A1.

3. Steve Yacino, "South Dakota Ranchers Face Storms Too, but U.S.' Helping Hands Are Tied," *New York Times*, October 16, 2013, A14; Adam Nagourney and Richard Pérez-Peña, "Shutdown's Pinch Leaves Governors with Tough Calls," *New York Times*, October 5, 2013, A10.

4. Jonathon Weisman and Jeremy Peters, "Republicans Back Down, Ending Crisis over Shutdown and Debt Limit," *New York Times*, October 17, 2013, A1; House vote 550, October 16, 2013, http://politics.nytimes.com/congress/votes/113/house/1/550.

5. Jerry Hagstrom, "Farm Bill Could Benefit from Shutdown Crisis," *National Journal*, October 6, 2013.

6. See Charles W. Johnson, *How Our Laws Are Made*, US House of Representatives, 108th Congress, 1st session, 2003, http://thomas.loc.gov/home/lawsmade.bysec/final.action.html.

7. Jerry Hagstrom, "Ready to Rumble over the Farm Bill," *National Journal*, October 20, 2013.

8. Rogers, "House-Senate Farm Bill Talks OK'd."

9. David Rogers, "Dem Effort to Relink Farms, Food Aid Fails Narrowly," *Politico*, October 12, 2013.

10. David Rogers, "Farm Talks Open with Optimism," *Politico*, October 30, 2013.

11. Jerry Hagstrom, "Farm Bill Conference Begins with Members Determined to 'Show How to Govern,'" *National Journal*, October 30, 2013.

12. Ibid.; Rogers, "Farm Talks Open with Optimism."

13. Jerry Hagstrom, "Compromise Is the Key to a New Farm Bill," *National Journal*, November 3, 2013.

14. David Rogers, "Corn Popping in Farm Bill Talks," *Politico*, November 19, 2013.

15. David Rogers, "Frank Lucas: This Week Key to Farm Deal," *Politico*, November 19, 2013.

16. Ralph M. Chite et al., *The 2013 Farm Bill: A Comparison of the Senate-Passed (S. 954) and House-Passed (H.R. 2642, H.R. 3102) Bills with Current Law* (Congressional Research Service, 2013).

17. Bill Tomson, "Farm Bureau: Don't Kill Old Farm Bills while Passing New One," *Politico*, October 16, 2013.

18. David Rogers, "Intense Lobbying Threatens Farm Bill," *Politico*, December 3, 2013.

19. David Rogers, "Farm Bill Talks Intensify," *Politico*, November 20, 2013; David Rogers, "Farm Bill Talks Stumble," *Politico*, November 21, 2013; Jerry Hagstrom, "Farm Bill Is Closer than Many Think," *National Journal*, December 1, 2013.

20. Rogers, "Farm Bill Talks Intensify."

21. Hagstrom, "Farm Bill Is Closer than Many Think."

22. David Rogers, "Big Trades Advance Farm Bill Talks," *Politico*, December 4, 2013.

23. David Rogers, "Farm Negotiators Shift on Subsidies," *Politico*, December 5, 2013.

24. David Rogers, "Reformers Could Be Crucial in Farm Bill Passage," *Politico*, December 9, 2013.

25. Bill Tomson, "White House Slams Food Stamp Cuts in Pre-Holiday Report," *Politico*, November 26, 2013; Ron Nixon, "Agreement Seen on Food Stamp Cuts in Farm Bill," *New York Times*, January 9, 2014.

26. Jill Lawrence, *Profiles in Negotiation: The 2014 Farm and Food Stamp Deal* (Center for Effective Public Management, Brookings Institution, 2015), 12.

27. David Rogers, "House Eying Short-Term Farm Extension," *Politico*, December 9, 2013.

28. Jerry Hagstrom, "GOP Has Much at Stake in Farm Bill," *National Journal*, January 5, 2014.

29. David Rogers, "No Farm Bill in 2013," *Politico*, December 10, 2013.

30. Jonathan Weisman and Jeremy Peters, "House Passes Budget Act and Military Abuse Protections, but Not Farm Bill," *New York Times*, December 13, 2013, A24.

31. David Rogers, "Farm Bill Optimism on CBO Numbers," *Politico*, December 12, 2013.

32. David Rogers, "Debbie Stabenow 'Feeling Very Good' about Farm Bill," *Politico*, January 7, 2014.

33. Ibid.; Erik Wasson, "Dairy Fight Could Hold up $1T Farm, Food Stamp Bill," *Hill*, January 8, 2014.

34. David Rogers, "Farm Bill in Trouble," *Politico*, January 9, 2014.

35. Brian Neely, "Dispute between Peterson and Boehner Part of Latest Farm Bill Snag," Minnesota Public Radio, January 10, 2014, http://blogs.mprnews.org/capitol-view/2014/01/dispute-between-peterson-and-boehner-part-of-latest-farm-bill-snag/.

36. Ron Nixon, "Farm Bill Compromise Will Reduce Spending and Change Programs," *New York Times*, January 27, 2014.

37. David Rogers, "Farm Bill Agreement Heading to Floor," *Politico*, January 27, 2014.

38. Ibid.

39. David Rogers, "New Farm Bill Readied for Final Debate," *Politico*, January 26, 2014.

40. Nixon, "Farm Bill Compromise Will Reduce Spending."

41. Rogers, "Farm Bill Agreement Heading to Floor."

42. Nixon, "Farm Bill Compromise Will Reduce Spending."

43. Ron Nixon, "House Approves Farm Bill, Ending a 2-Year Impasse," *New York Times*, January 30, 2014, A14.

44. Rogers, "Farm Bill Agreement Heading to Floor."

45. Ibid.

46. David Rogers, "House Poised for Vote on Farm Deal," *Politico*, January 28, 2014.

47. Nixon, "House Approves Farm Bill."

48. Rogers, "House Poised for Vote on Farm Deal"; Bill Tomson and Tarini Parti, "Plenty of Winners and Losers in New Farm Bill," *Politico*, January 28, 2014.

49. Rogers, "Farm Bill Agreement Heading to Floor."

50. Rogers, "House Poised for Vote on Farm Deal."

51. Erik Wasson, "House Passes $956B Farm Bill," *Hill*, January 29, 2014.

52. Ron Nixon, "Senate Passes Long-Stalled Farm Bill," *New York Times*, February 4, 2014.

53. Roll-call vote 31 on H.R. 2642, January 29, 2014, http://politics.nytimes.com/congress/votes/113/house/2/31.

54. Erik Wasson, "Ryan Explains Reversal on Farm Bill Vote," *Hill*, January 29, 2014.

55. Nixon, "House Approves Farm Bill."

56. Wasson, "House Passes $956B Farm Bill."

57. Jerry Hagstrom, "Why Is the Farm Bill Finally Ripe for Passage?" *National Journal*, February 2, 2014.

58. Ron Nixon and Derek Willis, "No Help for Farm Bill from Miffed Kansans in the House," *New York Times*, January 31, 2014.

59. Wasson, "House Passes $956B Farm Bill."

60. Ibid.

61. David Rogers, "Congress Approves 5-Year Farm Bill," *Politico*, February 4, 2014; Nixon, "Senate Passes Long-Stalled Farm Bill."

62. Michael Shear, "In Signing Farm Bill, Obama Extols Rural Growth," *New York Times*, February 7, 2014, A12.

63. Ibid.

64. Barack Obama, "Remarks of the President at Signing of the Farm Bill," White House, February 7, 2014, https://www.whitehouse.gov/the-press-office/2014/02/07/remarks-president-signing-farm-bill-mi.

65. Ibid.

66. Jerry Hagstrom, "At the Farm Bill Signing," *Agweek*, February 10, 2014.

67. Ibid.; Eric Wasson, "GOP Skips Obama Farm Bill Signing," *Hill*, February 7, 2014.

CHAPTER TEN. WHAT JUST HAPPENED HERE? FINDING MEANING IN THE POLITICS OF THE FARM BILL

Epigraph: Quoted in Bill Tomson, "Conaway: Ready to Work on Next Farm Bill, Immigration," *Politico*, November 19, 2014.

1. Barry Goodwin and Vincent Smith, "The 2014 Farm Bill—An Economic Welfare Disaster or Triumph?" *Choices* 29, 3 (2014): 1–4.

2. Jennifer Steinhauer, "Farm Bill Reflects Shifting American Menu and a Senator's Persistent Tilling," *New York Times*, March 9, 2014, A16.

3. E. Scott Adler and John Wilkerson, *Congress and the Politics of Problem Solving* (New York: Cambridge University Press, 2012).

4. Nadine Lehrer, "From Competition to National Security: Policy Change and Policy Stability in the 2008 Farm Bill" (Diss., University of Minnesota, 2008), 40–41.

5. Farm Aid still exists as a nonprofit organization to help small family farms and still holds annual concerts that feature John Mellencamp and Willie Nelson, but the events do not have their former visibility.

6. Quoted in Seth McLaughlin, "Democrats' Use of Farm Bill as Election Year Ammo May Have No Impact," *Washington Times*, September 25, 2014.

7. Editorial Board, "The Worst Voter Turnout in 72 Years," *New York Times*, November 12, 2014, A26.

8. McLaughlin, "Democrats' Use of Farm Bill."

9. Jennifer Jacobs, "Joni Ernst Wins Iowa U.S. Senate Seat," *Des Moines Register*, November 5, 2014.

10. David Rogers, "Farm Bill Foes Find Trouble of Their Own," *Politico*, September 10, 2014.

11. Ibid.; Jake Sherman, "How to Blow an Easy GOP Win," *Politico*, October 19, 2014; Karl Etters, "Gwen Graham Defeats Steve Southerland," *Tallahassee Democrat*, November 5, 2014; Alex Leary, "How First-Time Candidate Gwen Graham Defeated a GOP Incumbent," *Tampa Bay Times*, November 7, 2014.

12. Alex Leary, "Washington 'Outsider' Steve Southerland Steps through the Revolving Door and Joins Lobbying Firm," *Tampa Bay Times*, May 7, 2015; Amy Sherman and Mary Ellen Klas, "Another Graham for Governor? U.S. Rep. Gwen Graham Announces Potential Run," *Miami Herald*, April 21, 2016.

13. See Thomas P. O'Neill, *Man of the House:The Life and Political Memoirs of Speaker Tip O'Neill* (New York: Random House, 1987).

14. Robert Costa, "Republican House Majority Leader Eric Cantor Succumbs to Tea Party Challenger Dave Brat," *Washington Post*, June 11, 2014; Jonathan Martin, "Eric Cantor Defeated by David Brat, Tea Party Challenger, in G.O.P. Primary Upset," *New York Times*, June 11, 2014, A1.

15. Jerry Hagstrom, "Poised to Chair Senate Ag Committee, Roberts Pledges Not to Reopen Farm Bill," *National Journal*, November 11, 2014.

16. Rogers, "Farm Bill Foes Find Trouble."

17. Chase Purdy, "Stabenow to Stay on Ag Committee," *Politico*, November 17, 2014.

18. Scott Bomboy, "Why Boehner's Resignation Is Truly Historic for House Speakers," *Constitution Daily*, September 30, 2015, http://blog.constitutioncenter.org/2015/09/why-boehners-resignation-is-truly-historic-for-house-speakers/.

19. Emmarie Huettman, "G.O.P. Opposition to Gay Rights Provision Derails Spending Bill," *New York Times*, May 27, 2016, A13.

20. "Kansas Rep. Huelskamp Hit by Farmers' Backlash," *Kansas City Star*, July 31, 2014.

21. Justin Wingerter, "At Town Halls, Rep. Tim Huelskamp Is at Home," *Topeka Capital-Journal*, April 5, 2015. See also http://huelskamp.house.gov/about/events.

22. Justin Wingerter, "Rep. Tim Huelskamp's Dissertation: Agricultural Price Supports Ignore 'Economic Reality,'" *Topeka Capital-Journal*, May 14, 2016.

23. John Montgomery, "Huelskamp-Marshall Debate Reveals Clear Personality Difference," *Hutchinson News*, June 30, 2016; Josh Arnett, "Meeting Face-to-Face," *Dodge City Daily Globe*, July 1, 2016.

24. Ian Kullgren, "How the 'Big First' Primary Could Hinge on Farmers," *Politico*, August 3, 2016.

25. Timothy Carney, "How Believing in Free Enterprise—and Just Being Difficult—Led to a Republican's Defeat," *Washington Examiner*, August 4, 2016.

26. Catherine Ho, "Republican Primary in Kansas Highlights Continuing Battle between Tea Party, Establishment," *Washington Post*, July 29, 2016.

27. Curtis Tate, "Roger Marshall Upends Incumbent Tim Huelskamp in Kansas' 1st District GOP Primary," *Wichita Eagle*, August 2, 2016; Elena Schneider, "Huelskamp Loses GOP Primary Fight after Ideological Battle," *Politico*, August 2, 2016. On Boehner's reaction, see https://twitter.com/hillhulse/status/760668631923499008.

28. Rachel Blade, "Freedom Caucus Knives out for Ryan after Huelskamp Loss," *Politico*, August 3, 2016.

29. Justin Wingerter, "Inside Roger Marshall's Historic Win: How He Beat Rep. Tim Huelskamp," *Topeka Capital-Journal*, August 3, 2016; Blade, "Freedom Caucus Knives out for Ryan."

30. For my choice of section heading, apologies to Ross Parker, Hughie Charles, and Vera Lynn, but probably not Stanley Kubrick.

31. US Department of Agriculture, "USDA to Purchase Surplus Cheese for Food Banks and Families in Need, Continue to Assist Dairy Producers," press release no. 0181.16, August 23, 2016.

32. Jerry Hagstrom, "Farm Bill Debate Begins Early," *National Journal*, February 10, 2016.

33. Alan Bjerga, "Farmers Get Biggest Subsidy Checks in Decades as Prices Drop," *Bloomberg News*, April 11, 2016.

34. Vincent Smith, *A Midterm Review of the 2014 Farm Bill* (American Enterprise Institute, 2016).

35. Mary Claire Jalonick, "The Big Story: Only 4 States Will See Cuts to Food Stamps," Associated Press, September 17, 2014.

36. Robert Pear, "Thousands Could Lose Food Stamps as States Restore Pre-Recession Requirements," *New York Times*, April 2, 2016, A10.

37. Food and Nutrition Service, USDA, *SNAP Participation and Costs, 1969–2015*, May 16, 2016, http://www.fns.usda.gov/sites/default/files/pd/SNAPsummary.pdf.

38. Hagstrom, "Farm Bill Debate Begins Early"; Jerry Hagstrom, "The Future of Food Stamps," *National Journal*, April 28, 2016.

INDEX